MW01092957

HOW TO SURVIVE AGAINST THE ODDS

TALES AND TIPS FOR

ANIMAL ATTACKS AND

NATURAL DISASTERS

HOW TO SURVIVE AGAINST THE ODDS

WILLIAM MORROW

An Imprint of HarperCollins*Publishers*

CONTENTS

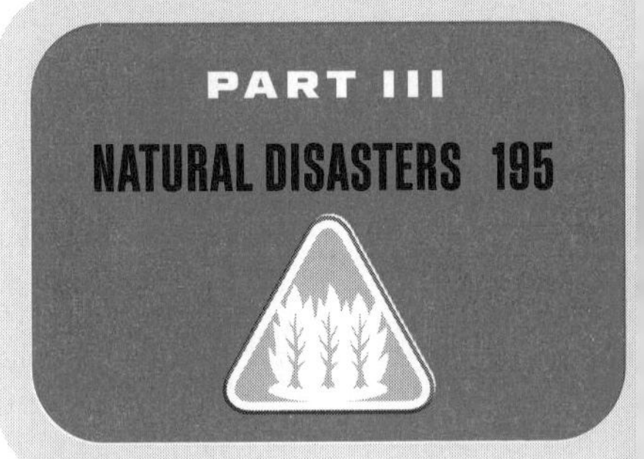

INTRODUCTION

We humans have a superpower, one that we often overlook. We may not be able to sprint as fast as a cheetah, or climb through the treetops like an orangutan, or dart through the ocean waves like a dolphin. But we have a much more subtle ability. Whether we're trekking through the icy tundra, lost in the oppressive heat of the desert, caught in a tropical cyclone, or adrift at sea . . .

We fight. We adapt. We problem-solve. And, ultimately, we survive.

When facing life-or-death nightmare scenarios, we discover our true resilience. Even when the odds seem insurmountable, even when hope seems lost, humans rise to the challenge. We may emerge battered, bruised, and shaken—but we triumph.

This book is rooted in that truth: that everyday humans (including you) can overcome the impossible. It's written in the spirit of *Against the Odds*, the hit podcast we cohost for . Our show takes listeners inside incredible tales of human survival, sharing moment-by-moment details about how, precisely, the heroes of our stories persevered.

As with our show, every story you're about to read is true. While we do take liberties with some dialogue used in the chapters, we stick to the facts of each event because those don't need dramatizing. They're harrowing enough.

But the stories aren't the only incredible part. These pages go beyond our podcast to explore how the human body is hardwired with an amazing array of mental and physiological tools that switch on by pure instinct when our lives are on the line. To understand these abilities—and their limits—we've consulted medical experts, who explain the fascinating science and inner workings of our bodies and minds during these extreme moments.

With the help of these experts, we explore real-life examples of what happens when humans are plunged into harsh, unyielding environments—from the Arctic to the Amazon rainforest—and how we fare in the face of natural disasters, such as city-engulfing wildfires or devastating tornadoes, as well as attacks from deadly predators, such as rattlesnakes and great white sharks. For fans of the *Against the Odds* podcast, many of these stories will be brand new—but we revisit a few familiar tales as well and use them to arm you with survival tips that could save your life.

Hosting *Against the Odds* has given us a profound appreciation for the strength of the human spirit. As adventure travelers and environmental advocates, we feel a strong connection through our own experiences to the stories of survival on our podcast and in this book. For Cassie, those challenges have included surviving twenty-one days naked in the Panamanian wilderness, with nothing but foraged food and water; competing in full- and half-Ironman triathlons across five continents; and plunging into the frigid waters of Antarctica. For Mike, it was buying a crocodile from a bushmeat seller after running out of food in the Congo rainforest, only to trade it later for a place to sleep, and finding a real crystal skull in the Black Hole of Belize. Each experience has deepened our understanding and appreciation of humanity's ability to survive in every kind of extreme situation the natural world can throw at us.

We hope you never find yourself face-to-snout with a snarling grizzly bear, or swept away in a flash flood, or fending off frostbite in an alpine blizzard. But should the worst happen, this book's trove of practical tips will help you defy the odds.

—MIKE COREY & CASSIE DE PECOL, HOSTS OF THE *AGAINST THE ODDS* PODCAST

PART I

EXTREME ENVIRONMENTS

The vast open ocean. Humid rainforest jungles. Unrelenting deserts. Frigid snowcapped mountains. Extreme locales around our planet are as beautiful as they are deadly. Adventurers and explorers in far-flung places face a variety of threats, from harsh climates to savage predators. Surviving in these environments requires careful preparation, quick thinking, a heap of resilience, and a bit of luck . . .

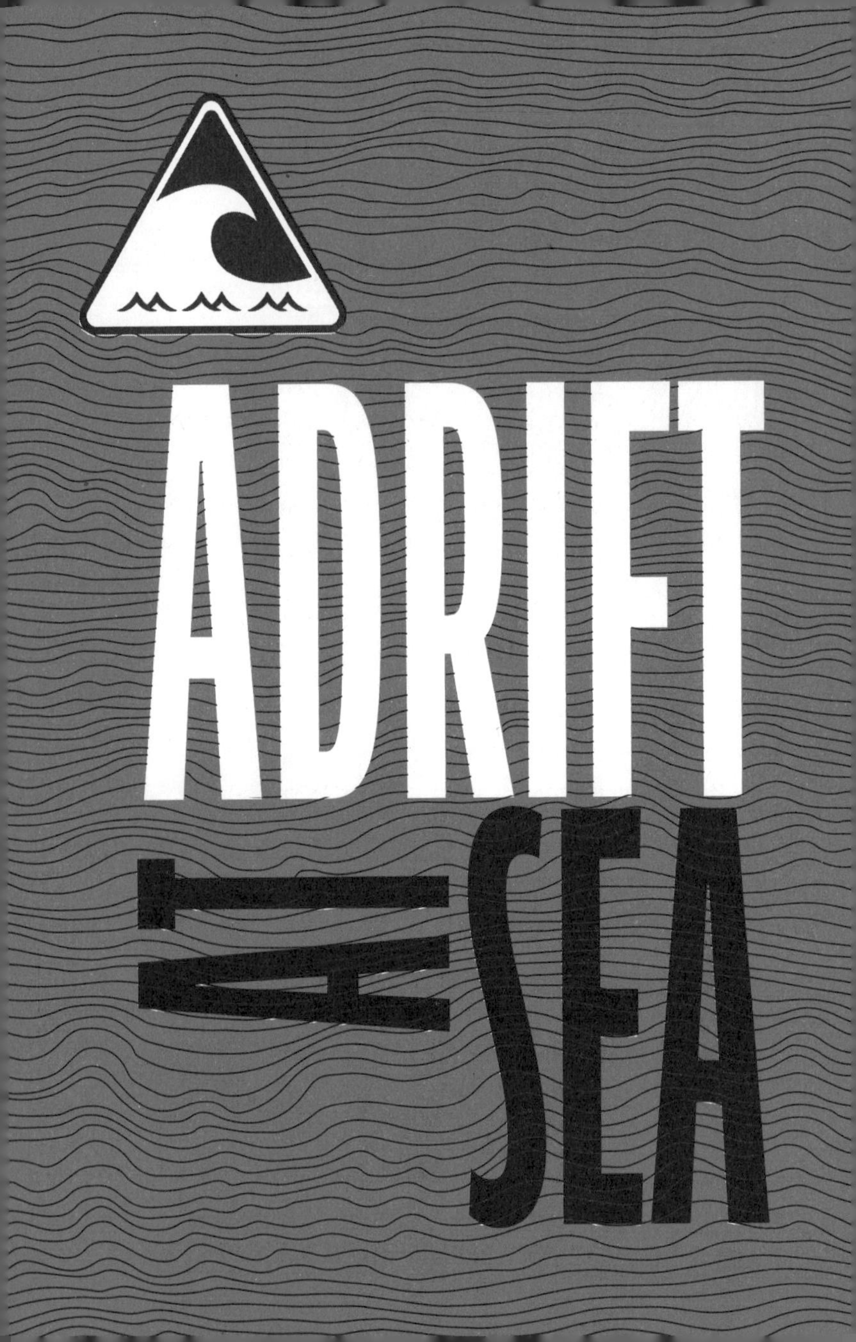
ADRIFT
AT
SEA

1

MARCH 4. MORNING.

Maralyn Bailey gazes out the porthole as she pours her morning coffee. It's seven o'clock and the sun's just cresting the horizon, casting a beautiful orange hue across the calm Pacific Ocean. She's a few hundred miles from Panama, bound for the Galápagos Islands. The thirty-two-year-old smiles as her thirty-foot sailing yacht bobs gently in the water.

Five years before, in 1968, Maralyn convinced her husband, Maurice, to sell everything, buy a boat, and leave behind their quiet suburban English life to sail around the world. They barely knew port from starboard, much less anything about piloting a boat across the ocean. Maralyn didn't even know how to swim. But after years of sailing lessons, in 1972, they set off from England, bound for New Zealand. They crossed the Atlantic, then sailed through the Panama Canal, entering the Pacific Ocean.

It's been nine months now since they left home, and Maralyn still has to pinch herself at their good fortune. Their boat, *Auralyn*—a combination of both their names—has been not just a sturdy vessel but also a cozy home. She glances around the cabin, full of the couple's books and small knickknacks they couldn't bear to part with. Everything is in its place; everything is perfect.

Maralyn pours a cup of coffee for Maurice, who's still asleep, and places the mugs on a small tray alongside toast and fresh mangoes from their last port, in Panama. Spending an hour on deck having breakfast with the love of her life is Maralyn's favorite ritual. She strokes Maurice's arm to rouse him.

"Morning, darling," she says. "Ready for breakfast?"

"Mmm," Maurice groans. "More sleep would do nicely. But I'll be right up."

Maralyn grabs the tray and heads up the narrow staircase to *Auralyn*'s deck. She's a few steps from the top when there's an enormous bang. The boat lists violently, and the tray slips from her hands and crashes to the deck. Maralyn cries out as she's hurled backward down the stairway. She lands on her tailbone in the cabin, her arms reaching back to lessen the impact.

Dazed, Maralyn looks over at Maurice. He's been flung from the bed and is lying on the floor, looking confused. He scrambles to his feet, and the couple run upstairs onto the deck and look out toward the sea. There, floating next to their boat, is a giant sperm whale, forty feet or maybe longer—far larger than *Auralyn*. The animal is close enough that the Baileys can see the scratches and barnacles on its dark gray skin. And something else: blood. A pool of crimson surrounds the creature, staining the ocean. It's growing by the second.

"It must've been harpooned and escaped," Maurice says. "No way we did that."

"Poor thing," Maralyn says. "What can we do for it?"

"Forget the whale—what about our boat!"

They hurry back down into the cabin. The compartment is already full of water up to their ankles. Maurice flings open the doors to the equipment locker in front of the cabin toward the boat's prow. What Maralyn sees terrifies her: a huge gash in the hull of the boat, eighteen inches long and twelve inches wide, just below the waterline. Seawater pours in, gallons every second.

As the water filling the cabin starts to lap at their calves, Maralyn has an idea. "Grab the bedding!" she shouts. They cram pillows and blankets into the gash, but it proves useless. They spend twenty more minutes trying to stanch the water's flow, but nothing works.

As the water reaches their knees, the realization hits Maralyn: *We have to abandon* Auralyn. *Or sink with it.*

* * *

It's been just thirty minutes since the impact, and the water in the cabin is already waist-deep. Maurice figures they have no more than an hour before the boat founders and sinks to the ocean floor. He doesn't have the tools or the time to fix the gash left by the whale. He and Maralyn move methodically through the cabin, gathering supplies. Thirty-two cans of food. Packages of cookies, nuts, dates, and a cake wrapped in plastic. Ten gallons of drinking water. A bucket and a trash can to catch rainwater. A first-aid kit. Six distress flares. They grab clothes and rainwear. They pack tools and equipment like binoculars, knives, paddles, a compass, charts, and a sextant—an instrument to determine location.

What they lack is a radio, or any other electronic transmitter for that matter. Maurice said he wanted to navigate by the stars, as great explorers had for ages. All their friends had said this was crazy. Now Maurice realizes they may have had a point.

Maurice eyes the pile of provisions that they've arranged on the deck. If they're careful, the rations should last twenty days, he figures. Maybe more. By his estimation, they are three hundred and fifty miles due east of the Galápagos Islands. He wipes sweat from his brow. It's still morning but it's already ninety degrees. It takes a half hour to do manually, but despite the commotion Maurice is able to pump up two inflatable life rafts. Now he and Maralyn load the provisions into them.

One's a tiny circular orange raft, no bigger than a two-person tent, with a small funnel-shaped canopy to protect them; the second is a gray rubber rowboat, barely big enough to hold one person. That will carry most of their supplies. Both rafts look paltry compared to their sailing yacht, and even smaller against the vastness of the sea that surrounds them.

As they shove off from *Auralyn*, Maralyn sobs uncontrollably. The boat had meant everything to her. Maurice squeezes her hand while she weeps, and the two watch in silence as *Auralyn* slips beneath the surface, barely an hour after the whale struck it.

Between tears, Maralyn asks Maurice about their chances of survival. What can he say? He has to stay positive—for both their sakes. It could be weeks before they're reported missing—they're not expected in the Galápagos until then. But they are near a major shipping route, so Maurice says that odds are they'll be spotted sooner or later. Maralyn takes a few deep breaths and tightens the strap on her life jacket. Shc still doesn't know how to swim. They agree to three-hour shifts: One will rest while the other looks for passing ships. This is their routine for the next two days.

On the third day, they start rowing. Using the sextant and charts, Maurice estimates that they're one hundred miles north of the Humboldt Current, which could push them to the Galápagos in less than two weeks. Maurice reckons that,

to allow for all that time drifting, they must reach the current in ten days, otherwise their food supplies will run out before they reach land. That means they must row ten miles per night—it's too hot during the day.

Privately, Maurice doubts they can cover that much distance. Their hands quickly become so blistered from rowing that they have to pause every few hours, and the dinghy, laden with supplies, is a drag on their progress.

But they have to try. So for three nights, they row incessantly. On the sixth day after losing *Auralyn*, Maurice checks the sextant to find they've moved just ten miles toward the current. No matter how hard they row, the ocean is pulling them in the opposite direction. He breaks the news to Maralyn.

"We can't make the Humboldt Current," he says. "We don't have the water or food—or the strength—to keep up this pace."

Now it's Maralyn's turn to be positive. She suggests they stop rowing and let the ocean push them toward another current. Surely there are other shipping lanes, she says.

Maurice nods, but secretly he's worried. All the effort from rowing has already burned up the majority of their food and water. He looks at the remaining rations. Only enough for five more days.

And they're about to drift farther into the North Pacific. Farther from land. Farther from humanity.

*   *   *

DAY 9. Just after daybreak, Maralyn shifts uncomfortably in the life raft. Pressure sores on her backside make sitting painful. She looks at Maurice, who also grimaces. His sores are worse. There's simply no room to move, and it's reducing their blood flow.

> **"In the case of pressure sores, the pain may diminish if the wound depth is below the level of sensory nerves," says Beth Palmisano, a medical doctor who specializes in pain management. "A pressure sore may be less painful if there's nerve damage due to trauma. In those cases, the wound is so deep that nerve endings aren't being stimulated anymore, so it may hurt less."**

To keep her spirits up, Maralyn's determined to keep their breakfast routine. She passes her husband a plate of meager rations: half a date, some cookie crumbles, and a tiny cup of water. As they chew, she gazes out of the canopy's opening into the endless blue that surrounds them. This view once brought her such pleasure; now she wonders if it'll be the last thing she sees.

On the horizon, something shimmers. *Is that . . . a ship?* Maralyn leaps to her feet, grabbing the binoculars. It's a small white fishing boat.

"Maurice, we're saved!" she shouts and jumps, rocking the life raft. The boat's at least a mile away. She shoves a flare into Maurice's hand, and he tears off the cap and strikes it. Nothing happens. He tries again. And again. It's a dud. Screaming, he throws it into the ocean while Maralyn grabs another flare.

This one fires into the sky. *Please see it.* Maralyn's eyes are locked on the boat. *Please.* The boat doesn't change course. And it's starting to recede in the distance. It's headed away from them.

"No! Light another one!" Maralyn barks. Maurice fires a second flare. They scream until they're hoarse. But the ship fades into the horizon. Once again, they're alone, and Maralyn feels helpless at keeping the dark thoughts from her mind. How will they die? From thirst? A shark attack?

A bump from below interrupts Maralyn's thinking. Another bump, hard enough to knock over a stack of food tins. Maralyn pokes her head out the canopy's small window.

Four turtles are hitting the boat. The sea turtles ram the life raft again. *You want to hunt us? We can hunt you.* She plunges her hand into the sea and grabs one of the shells. She hoists the animal up and turns to Maurice, handing him a paddle.

"Knock it out first," she says. "Then I'll cut off its head."

The notion of murdering turtles doesn't sit well with Maurice; Maralyn can tell from the pained look on his face. But they're nearly out of food. He nods and complies.

Soon enough, Maralyn's prepared a bowl of finely chopped raw turtle meat. They take a reluctant bite; it's chewy but not bad. It's the first fresh food they've had in more than a week. They finish the meat quickly.

* * *

DAY 40. For Maurice, the days have become a blur. There'd been a second ship sighting, about two weeks ago. He and Maralyn set off two flares, to no avail. And four days ago, they'd used the last flare on another passing ship. Again, no luck. Maurice keeps kicking himself for not including a signal mirror in their gear.

On the plus side, massive rainstorms rolled in just as they were on the last gallon of drinking water. They collected it in the bucket and the trash can and now have plenty. Maralyn fashioned a fishing hook from a safety pin and tied it to some thread. It turns out turtle meat is quite effective bait. They have a few cans of food left but are mostly living off raw fish and turtles, which keep surfacing near their raft. Turtle meat is even growing on Maurice; it tastes like a cross between chicken and lobster, with a bit of crab mixed in.

But he feels horrible. His vision keeps blurring. He has a fever and a bad earache. He lies on his back in the dinghy and looks up at Maralyn, who's cleaning a fish. Her face is gaunt, eyes sunken; her once graceful body is now so thin, he can see her ribs. He wonders how much longer they have, drifting out here. It feels like death is toying with them, waiting for them to give up.

> **"When the sun hits the sea, it's reflected at greater intensity. This is damaging to the eyes. Chronic and prolonged exposure to sunlight reflected off the sea can cause pterygium—tissue growth in the eye that can lead to blindness."**

* * *

DAY 94. It's raining so hard, Maurice can hardly see. It's been like this for a week. But they have to eat. He slides from the covered life raft to the rubber dinghy, still connected by the same several feet of rope that's tethered the raft and the dinghy together for over three months. He throws a hook baited with turtle meat into the raging ocean.

Amid bursts of lightning and claps of thunder, Maurice squints against the sideways rain. Violent waves roll across the ocean's surface, great swells churning in all directions. Maralyn's head pokes out of the life raft's lookout window, checking on him. He gives a feeble wave.

Then he freezes. Behind her a rogue wave is rising. It's about to crush them.

Maurice screams, but the roar of water drowns him out. His stomach drops as he watches Maralyn and the life raft shoot skyward. Then the wave comes for Maurice.

Cold water slams him down onto the dinghy's inflatable floor. Then the whole boat is underwater, sinking. Maurice debates opening his mouth, just to end this terrible ordeal. But an image of Maralyn tossed into the ocean pops into his mind. *She can't swim. Keep fighting. Go help her.*

His head pops above the surface, and he takes a huge gulp of air. He gets a glimpse of the life raft, fifty feet away. Another wave takes Maurice back under. Every few seconds, he surfaces, gulps a breath, and prepares for the next swell. He wonders when this horrible cycle will relent.

The next time his head pops up, he bumps into something. It's the dinghy, upside down. He clings to it, then feels around for the rope and follows it back to the life raft.

Maralyn's face comes into view. She's still in the life raft, and okay.

It takes all his energy, but he lifts himself into the raft and collapses onto the floor. Maralyn wraps her arms around him. His lips feel numb and he's trembling uncontrollably. But he feels safe in her arms, even as the storm rages on around them.

> Exposure to water under ninety-four degrees for a sustained period of time will induce hypothermia, as Maurice experienced after being hit by the rogue wave. All boat crash survivors, particularly those in open craft, are likely to get hypothermia. Cold nights can induce hypothermia too.

* * *

DAY 109. Maurice is running a high fever and suffering from a nasty cold. The temperature swings aren't helping much.

> Sunburn can be devastating. Even just a few hours in the sun can cause second-degree burns. Sunburns can cause headaches, nausea, and fatigue, if severe. "Redness on skin is a first-degree burn. Secondary burn, you're going into the next layer of the skin," says Dr. Deepak Sachdeva, a medical doctor who specializes in emergency medicine. "This is much worse, where you get blisters." The skin is your major protective barrier from the outside world and keeps you from getting infections. "When you get blisters, you're losing protection, and bacteria and germs can get in there," says Sachdeva.

During the day, it's too hot to wear clothes; at night it feels like they may freeze to death. Maurice lies against the life raft's canopy and struggles against nodding off.

> In tropical climates, as the Baileys experienced, prolonged exposure to the sun and heat can bring on severe cases of sunburn and hyperthermia, or heatstroke. When your body can't cool down quickly enough, you start to overheat.
>
> "As your core temperature rises, you'll get heat exhaustion first," says Sachdeva. "You may or may not experience syncope, which is passing out, and then heatstroke, the most severe. This is when your brain starts frying, and that's fairly irreversible. Once you get to the point where your body temp is so high, you have cell damage from the heat. You get confused and, in severe cases, your brain swells. If it swells enough, you'll have seizures and then fall into a coma—then death."

* * *

Maurice looks at Maralyn. She must've lost thirty pounds—but not her optimism. She conjures up mental images of fancy meals and shares every detail of the fictitious menus with Maurice. Her spirit keeps him going. But he knows now it's only a matter of time. He wonders if they'll ever have real food again. The last check of the sextant was disappointing. They are drifting northwest, now a thousand miles from the nearest coast. They are farther from land than they've ever been.

Maralyn's optimism in this bleak setting may seem unusual. One's ability to remain hopeful stems from a mix of nature (i.e., our genetics) and nurture. "What makes a profound difference is the environment in which you were raised," says Anthony Giovanone, a doctor of osteopathy and a psychiatrist. "When you're young, if one caregiver makes you feel safe, then you'll be more resilient, which enhances optimism. If no caregivers give you that feeling, it can lead to increased pessimism. And, later in life, when you sense that things aren't okay, you're more likely to stay in a state of fear."

Maralyn drags a finger along the ocean's surface. "There's a little shark here. I think it wants me to pet it," she says. Sharks had been swimming around the raft for a while. It was scary at first, but they never attacked the boat, so the couple had gotten used to them. Maralyn's hand suddenly dives into the water.

It comes up holding a two-and-a-half-foot shark, writhing and snapping its sharp teeth.

The two pin the thrashing shark on the floor of the raft. Maurice wraps a towel around its jaws, and they hold it there until it dies. Maurice brings the dead shark to the dinghy to butcher it with a dull pocketknife. He's only made the first cut when Maralyn shouts behind him: "I've got another!" *What?* He pulls the dinghy close to the raft and sees Maralyn holding another shark by the tail. He takes it and stabs the shark in the gills, over and over. He's focused on killing this thing when Maralyn yelps again. "Another! I got another!"

> What makes the average person suddenly grab a two-foot shark by hand? "Habituation," says Giovanone. "The more you see something, the less of a fear response you have." Take firefighters, he says. Firefighters see flames so often, it takes away that fear response, allowing them to run into burning buildings. "When Maralyn's trying to pet the shark, her mind has determined that they're not dangerous; these sharks are always there. The perceived threat level goes down and you can do things you never would've before."

"For God's sake, please stop! I'm begging you," he exclaims.

Later, their bellies are full of shark meat and their mood lifts. Maurice is extremely weak, and severe pains rack his chest. But as they slowly fall asleep, they talk about where they'll sail to next after being rescued.

* * *

DAY 118. Maralyn looks at Maurice sleeping beside her. His breathing is quick and shallow. His health is deteriorating, she thinks. *He doesn't have much time left.*

> **Being sedentary in a cramped life raft can cause blood clots. When walking, your muscles cause compression to allow blood to flow back to the heart. "When you're unable to move, as this couple was, there's little to push the blood back," says Sachdeva. "When blood is stagnant, you risk developing blood clots. They develop in the legs first, and cause pain and swelling. If they break off and move to your lungs, they can cause pain to your chest, trouble breathing, or both. This can make you pass out and can even kill you." Sachdeva suggests making concerted efforts to move your legs every few hours, to activate your blood flow.**

But then—a low growl. Her head darts out the life raft's opening. The horizon's empty, but there's definitely a noise. A ship's engine! Maralyn wakes Maurice, then crawls into the dinghy. She waves a yellow rain jacket wildly, nearly tipping the boat, looking through binoculars.

Finally, from the east, a small white rusty fishing vessel appears. Maralyn jumps now, swinging the jacket in the air like a flag. Maurice joins in with another rain slicker. It's the first boat they've seen in forty-three days, and Maralyn refuses to give up. It looks like it'll pass about a half mile from them. She screams until her lungs hurt.

But the ship looks to be changing bearing, heading away from the distressed couple. Maurice crumbles to his knees, deflated. Maralyn won't stop yelling and waving. Minutes feel like an eternity, but she doesn't relent. Finally, just before it disappears from view in the binoculars, Maralyn sees the ship turn.

It's coming back.

Maralyn clambers into the lift raft, grinning. "Come on, let's put some clothes on. We can't get rescued like this," she says. Maurice, dazed, looks down and remembers they're both completely naked. Maralyn laughs and tosses him a pair of shorts.

It's finally over.

* * *

South Korean fishing vessel *Weolmi 306* saw Maralyn's raincoat from more than a mile away. On board, they were fed, received medical attention for sores, and were given anti-inflammatory pills. Walking took days; their feet and legs swelled immensely each time they tried. It took two weeks to reach the nearest possible port in Honolulu.

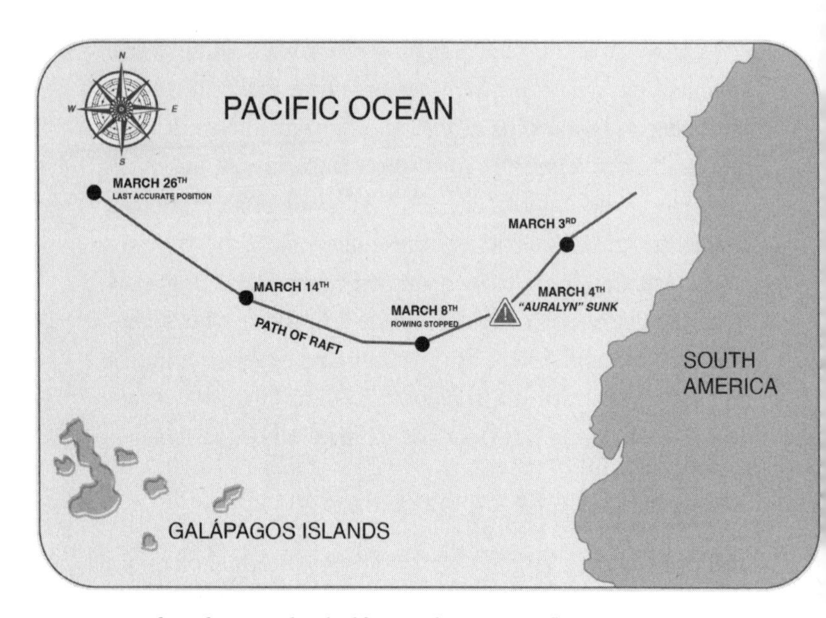

After four and a half months at sea, floating more than fifteen hundred miles, Maralyn and Maurice Bailey finally reached land.

Each had lost forty pounds. Maurice had a severe infection that left him nearly deaf in one ear. He also suffered thrombosis and a blood clot in his lungs, which caused permanent damage and breathing problems. Apart from the sores on her legs and backside, Maralyn had no serious issues or lingering ailments.

In 1974, one year after their rescue, the Baileys set sail in their new yacht, *Auralyn 2*.

HOW TO SURVIVE AT SEA

In 1972, the gash in *Auralyn*'s hull likely doomed the yacht to sink, no matter what the Baileys did. Today, that's not the case. Underwater epoxies and other sealing compounds and tapes can repair small breaches—at least enough to make it to the nearest port. Always have one of these kits on board. Your goal should always be to stay with your vessel, but if you're forced to leave, here's what to do next.

SEND A DISTRESS CALL. Don't do what Maurice did—always bring a radio. Before abandoning your ship, if your communication devices work, send your coordinates immediately. Send an SOS signal on your VHF radio, using Channel 16, the universal channel for emergency calls.

PACK ESSENTIAL GEAR. Bring all the potable water and canned goods you can, life jackets, clothes, buckets or plastic tarps to collect rainwater, tools to open cans and cut food, a first-aid kit, a bailing bucket, and signaling devices including flares, whistles, and mirrors.

USE A LIFE RAFT. Survivors with life rafts can survive for one hundred or more days, as the Baileys did. Nearly all modern life rafts have canopies, but build one if yours does not. You need protection from the sun. Avoid being in the water at all costs. Swimming drains your energy and increases dehydration; hypothermia can be quick to take hold, even in tropical oceans. Being in the water also exposes you to predators, such as sharks.

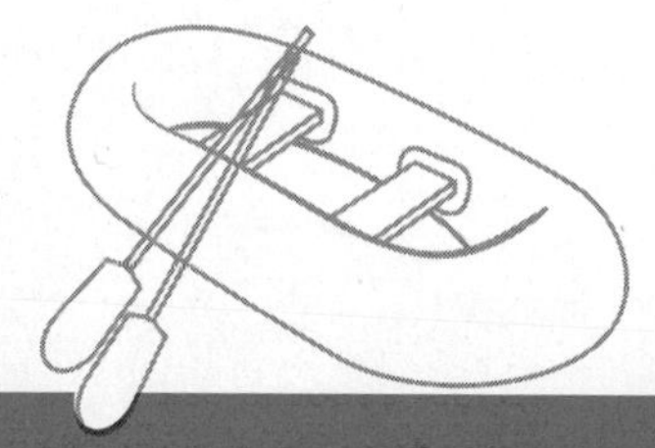

EMPLOY SEA ANCHORS. You want to stay close to your last location, so use a sea anchor or drogue to slow the rate of your drifting. These anchors also steady the raft in storms or rougher seas.

ONLY DRINK POTABLE WATER. Consuming seawater causes further dehydration and delirium, and it will kill you. Don't drink alcohol or the blood of birds or turtles; both blood and alcohol will dehydrate you further. Collect rainwater in clean containers or build a solar still by placing a plastic tarp over seawater and letting the condensation run off. You need one liter of fresh water a day to survive; any less and you'll start to feel dehydration symptoms like fatigue and headaches.

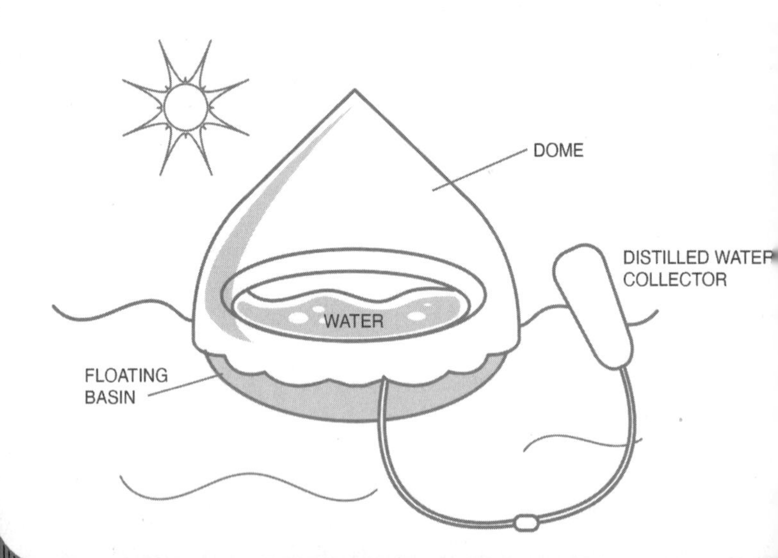

CLEAN YOUR SKIN AND MOUTH. Exposure to salt and dirt can cause infections. Avoid wearing seawater-soaked clothes for too long. Rain can be used to help bathe. If you have soap, use it.

EAT WHATEVER YOU CAN. Fish, turtles, and sharks often huddle under boats and life rafts. Fast hands can grab them. Otherwise, fishing lines and hooks are included in modern life rafts. At first, you can use anything sparkly as a lure; once you've succeeded in a catch or two, fish or turtle innards make great future bait. Try drying anything caught in the sun for a few hours—it'll improve the taste.

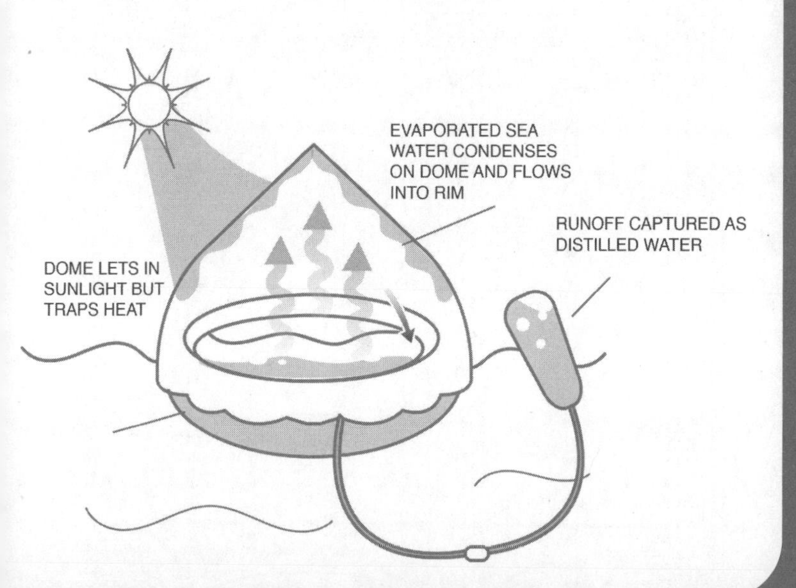

ESCAPING THE DESERT

SEPTEMBER 24. NOON.

Ed Rosenthal leaves his motel room and draws a deep breath of desert air. It's perfect. The sixty-four-year-old Los Angeles real estate broker just closed a momentous deal and is off to celebrate in his favorite manner: a hike through California's Joshua Tree National Park. Back in the 1980s, Rosenthal fell in love with desert hikes, and he has gone just about every weekend since. He loves the wide-open spaces and the solitude. There's something about the desert that revives his soul.

Today's route is a familiar one—a three-hour, five-mile loop from Black Rock Canyon to Warren Peak. He's done this journey ten times and likes coming across familiar landmarks, ones that have special meaning—like the rocky outcrop where he was when his daughter Hilary called, ecstatic at passing her driver's test. He grins at the recollection as he drives, opening the window to feel the heat from the sunny fall Friday. He pulls into the parking lot at the trailhead and looks at his watch. It's twelve thirty, and already ninety degrees. It's going to be a scorcher.

An older guy camping nearby gives a friendly wave as Rosenthal assembles his backpack. In goes a peanut butter sandwich, a few energy bars, dates, a small medical kit, rope, a space blanket, a headlamp, matches, and emergency flares. He also has about three liters of water in his CamelBak pouch, one for each hour of hiking. He considers bringing extra clothes, as well as camping tarps and extra water, but decides against it. All that extra gear just adds weight. Besides, he'll be back by three thirty.

He grabs his trusty walking stick, shoulders his bag, and winds his way up the trail. He's wearing shorts and a loose-fitting T-shirt. His climb is steady, with 1,100 feet of elevation to tackle. The path tightens in a few spots, tall rock faces and craggy pine trees looming overhead, but it's relatively easygoing.

After an hour and a half, he's enjoying the view from the top of Warren Peak, 5,100 feet above sea level. Between bites of his sandwich, Rosenthal sits on an outcropping and soaks in the majesty of the 10,834-foot San Jacinto Peak and the undulating terrain of gray boulder piles, sandy slopes, olive-green scrub, and gnarled Joshua trees. He closes his eyes, letting the warm winds wash over him, blowing away all thoughts except for living in this moment. Completely alone, he feels the opposite of lonely.

After thirty minutes, contented and full, he rises. It's time to head back. Rosenthal looks for the tiny trail he took on the way up, but suddenly he can't find it. The trees look eerily similar, as do all the boulders from his vantage point at the peak. His heart rate jumps as he spins around. *This is silly*, he thinks. *Just find your footprints and follow those.* But the harder he looks, the more unfamiliar everything seems.

There's no trace of his steps. He can't even find the trail marker he passed earlier.

He clambers a little way downhill, then blazes a trail around the peak, circling it to look for any trace of the path back down. Sweat seeps through his floppy sun hat, partially from the heat but also from a growing sense of panic. He can see for miles from this height, across the rolling hills below him. But he still can't find the right trail to lead him back to his car. It's been twenty minutes. If only he recognized

something, anything. He scans the area, looking for flashes of colored clothing. Perhaps another hiker in the area to follow.

There's no one.

Finally, Rosenthal spots a narrow chute between a pair of massive boulders. Beyond it, there's a six-foot drop into an arroyo, a water-carved channel. It heads toward where he believes he needs to go—down. *Down is good. We like down.*

He drops into the dry streambed, landing on a steep slope. He struggles on the loose soil, bracing with his walking stick as he sidesteps thorny cacti and loose rocks. This is not a trail, but he continues, picking his way carefully over a half mile. It seems to take forever, and then he finds himself standing on another precipice. He looks down—it's at least fifteen feet to the bottom, where craggy rocks stare up at him like dinosaur teeth.

Tugging the rope from his pack, he scans the area. If he can tie the rope around something sturdy, maybe he can rappel down. Then he thinks better of it. If the knot doesn't hold, he's done. Gingerly, he creeps to the edge, draws a steadying deep breath, and jumps down to a narrow ledge. Rocks and brush scrape his legs as he lands. A few more careful drops, and he's at the bottom. Banged up and breathless, barely standing against this harrowing incline, he looks up from where he came. There's no going back.

He's downhill from the peak, but to advance he's got to get over a hill. He saw this hill when he started down but it's much bigger standing beneath it. He clambers up, hundreds of feet, every inch littered with scrub bush and cacti. The spiny thorns dig into his exposed legs and arms. But he calmly continues. He's able to push away the panic because he's sure this'll dump him onto an actual trail. By some miracle, it does. He stumbles onto a narrow trail, without markers, heading downhill.

For the first time in the hour since he started back down the mountain, he feels his pulse slow down even as his pace quickens. He squeezes some water from his CamelBak into his mouth. A little longer and he'll be back at his car, chuckling about this.

* * *

Three hours later Ed Rosenthal is hopelessly lost.

He's still on the same narrow trail, hoping it will lead to a larger one, a path back to civilization. But miles have passed without delivering on that hope. No turnoffs, just unending trail. Yellow and purple prickly pear blooms dot the sides of the path, but around each bend is just more bleak desert.

Shimmering heat waves rise from the folded terrain, but the desert is also flattening ahead. Rosenthal suspects he is on the border of the Colorado Desert, having left the Mojave. Hotter and more unforgiving, the Colorado Desert is no place to be lost.

He clamps his teeth down over the nozzle that draws water from his CamelBak—and realizes it's empty. It's nearly six hours since his hike began, and the sun is dipping low across the desert. It'll be night soon. Pulling out his cell phone, he tries 911 again. Still no service, and his battery is close to dead.

> **Human bodies are made up of between 50 and 70 percent water, and every day our bodies need to replace about three liters of water that we lose through breathing, urination, and sweating. Once water levels drop, even as little as 2 percent, mild dehydration occurs. This causes your brain to contract, moving away from your skull and giving you a headache. And without water, your blood thickens, making it harder for nutrients and oxygen to reach your vital organs, leaving you fatigued.**

As the sun fades, Rosenthal accepts that this will be where he's sleeping tonight. He's too tired to continue, so he makes his way into a compact canyon, forty by sixty feet. He'll bed down here tonight. He plops down on the dusty earth and stares at the darkening sky. Airplane lights blink miles above him. He tries signaling by flashing his headlamp on the shiny side of his space blanket and blowing a whistle. It's fruitless.

He shoves his backpack under his head. It makes for a lumpy pillow, but it's better than a rock. He's wishing now he'd packed that tarp and those extra layers of clothes. The desert at night gets cold. But at least he has the space blanket. He dabs at his bleeding arms and legs with ointment from his first-aid kit, applying bandages over the most severe cuts. Sighing, he scratches his dry tongue, which feels like it's coated with sand, and closes his eyes.

No one knows he's lost; he wasn't due to call his wife, Nicole, until Monday. Can he really survive two more days in this hostile land?

* * *

When Rosenthal wakes up, around four in the morning on Saturday, it's already getting warm. He didn't sleep well, but he's determined to use what little energy he's regained to soldier on. In the waning moonlight, he packs up and picks his way up a pile of sandstone debris, some thirty feet high, emerging into a canyon he'd ambled around the day before.

Dozens of trails spider off in all directions. He selects one that appears to have been made by humans and sets off. He swallows the last of his one remaining Clif Bar, but it only increases his thirst, sticking to his mouth. This trail fades off to where it's no longer obvious, so he trudges back to the canyon and chooses another. He spends hours in this trial-and-error method, each time forced to backtrack to the canyon.

Sunrise brings scorching heat, more than one hundred degrees. Rosenthal knows it'll quickly zap his dwindling energy reserves. Hunkering down under a nearby pine tree, he spends the remainder of the day here. Every few minutes, he has to move out of the sun's searing rays. It's an exhausting, endless dance.

> Prolonged exposure to heat sets off a chain reaction of maladies, including a fast, strong pulse, headaches, dizziness, nausea, and confusion. "Our bodies are designed to function in a narrow temperature window," says Matt Cummins, a medical doctor and emergency physician. "As your core temp rises, your breathing and metabolic rate go up. The body compensates by pushing more blood to the skin, to help more heat evaporate there. But as you pull the blood from your core, you have issues with oxygen delivery to organs," says Cummins.
>
> The bowels and kidneys stop functioning first, and you'll have stomach pain as this happens. Your liver is the next to shut down, and this will throw off enzyme levels everywhere, including the brain, leading to increased confusion. "When your core temp hits one hundred and five degrees, cellular metabolism stops," says Cummins. "You're no longer able to create or use energy, and multi-organ failure happens. It's not a good way to die; it's slow and painful."

His thirst is unbearable, consuming every thought. He tries sucking on pebbles, which helps, but not enough. It's been twenty-four hours since his last sip of water and the only liquid available is his urine, which he's been passing into a cup, praying it wouldn't come to this. He studies the dark orange sludge. The stench is overwhelming as he lifts it to his face. Clenching his eyes shut, he shudders and takes in a mouthful. Immediately he gags and spits it out.

> **Under extreme heat scenarios, as Rosenthal faced, the body begins to make executive decisions about where to get water from internally. It first reduces nonessential reserves, like your kidneys. This is why your urine darkens as you get dehydrated. The more your organs shut down, particularly your kidneys, the faster toxins build up too. Next, the body starts redistributing water from other organs to keep your blood pressure up. Your gums recede, your eyes shrink, your lips dry up, and your tongue hardens. Under the most extreme dehydration, you won't bleed from a wound.**

Dusk brings relief from the heat. He crawls to a nearby patch of yucca plants, slicing some tendrils with his Swiss Army knife and sucking on them, desperate for any drops of water. Nothing. Angrily, he flings them away.

And then night comes, and with it the cold. Rosenthal shivers. He strikes a match over a pile of twigs, but between his trembling hands and the stiff desert winds, he can't get the flames to catch. He tries one of his emergency flares too, but it doesn't meaningfully catch, and fizzles. So do his hopes.

No one will be looking for him yet, his water's gone, and he's exhausted. Worse, the emergency blanket is disintegrating, with pieces blowing off as he covers himself. Mylar blankets are strong for how thin they are, but as soon as they're punctured, they easily tear to shreds. He tries to compensate by wrapping a meager roll of toilet paper around his body—he looks like a mummy. It offers little warmth and protection. Rosenthal knows that this will be a brutal night. He tracks the Orion and Vega constellations for hours, until exhaustion wins and he drifts off.

* * *

Sunday dawns and Rosenthal's outlook is bleak. He's been lost for forty-eight hours and knows he's too weak and dehydrated to walk out of here. Moving twenty feet just to crawl away from the sun takes all his energy. But he might be able to make it to a canyon a few miles away where it looks like there's shade.

By midmorning, he's only traveled one mile before the beating sun drives him to the shadow of a boulder. He had a quadruple bypass a decade ago, the result of a heart attack during a disastrous real estate deal, and he's worried about another heart attack.

Famished, he chews on a handful of dates, the last of his food. But his mouth is so dry it's like he's trying to swallow glue. With a mighty effort, he spits them out and leans back with a groan against the rock. The next eight hours are a vicious cycle of sleep, waking when the sun burns him, then crawling again to the shade. Snooze, scorch, scoot. Over and over.

> Without food, the body goes into starvation mode, or ketosis. "Your liver is the biggest storage of sugar that's used in rapid-need situations, when you don't have enough time to break down fat and muscle," says Cummins. "Once the sugar reserves are depleted, it's easier for the body to break down muscle than fat. From the muscle, your body is getting proteins and sugars. You're eating yourself, basically."

Come dusk, he's able to stagger the few miles to the canyon, but his body is giving out. His head pounds, an early sign of dehydration, and there's so little water left in his body that he's stopped sweating. He's incredibly hot and

can't do much about it. *Either I die here or they find me here. Maybe both.*

> **Effects of dehydration can come on quickly, depending on your body weight, the outside temperature, and your exertion levels. "In warm temperatures like Ed experienced, you can sweat up to two liters per hour," says Cummins. "That's a lot to lose if you don't have any water going in. When you're severely sunburned—actual burns or blisters on your skin—you lose the ability to hold water in too. Your body will do everything to compensate—the kidneys are great at keeping you at a net water balance between what's coming in and what you're losing—but only to a certain point. Three or four days of temperatures like Ed endured are about the maximum of what the body can handle. That's why he was basically lying on the ground." The average person can survive three days without water, though some can last longer.**

* * *

For the next seventy-two agonizing hours, Rosenthal crawls around the canyon during the daylight hours, playing a perverse game of hide-and-seek with the sun. He again tries sucking water from yucca and cacti, but it's useless. At night, he uses all his matches and flares but can't get a meaningful fire going. He drifts off in fitful spells, his body shaking in the cold. He worries about not being able to fend off hypothermia.

On Wednesday, he fishes a permanent marker from his pack and scribbles notes on his floppy hat, messages to loved ones, for when they find his remains. On the fabric, he writes poems to his wife and daughter, and instructions for his wake: Everyone should get drunk and stoned, and there should be a poet, and Persian food. He recites the Shema Yisrael, a Jewish prayer invoked at the end of one's life. Then he prays for rain.

His prayers are answered. That afternoon, rain falls. It's just a drizzle, but he flops onto his back, opens his mouth, and tries to catch every drop. It's not enough to slake his thirst, but it makes him smile for the first time in days.

Thursday morning is the first time Rosenthal isn't woken by the sun and heat. His eyelids won't open, and his body's not responding. *Death won't be long now*. Then a faint buzzing in the distance hits his ears. Its rhythmic thumping grows closer. Willing his eyes open, he sees a helicopter hovering overhead.

> **"The longer you go without food, the worse your brain performs, leading to sleeping issues and a lack of concentration. In extreme starvation, your body will resort to using heart and other vital organ tissue as fuel, leading to death. Still, most people can survive for weeks without food. "Severe hunger will pass," says Cummins. "When humans were cave dwellers, we sometimes didn't eat for days after failed hunts and still managed to survive.""**

"Hey! Are you that Rosenthal guy?" booms a voice from the copter's loudspeaker. He weakly gives a thumbs-up, all he can muster. The chopper finds a spot nearby and lands, sending a wash of dust and dirt over Rosenthal's immobile body.

He doesn't care. For the first time in six days, he knows he's going to live.

* * *

Ed Rosenthal was reported missing by his motel late Saturday, then again on Sunday morning by the man Rosenthal had passed entering the park, who noticed that Rosenthal's car hadn't moved. An enormous search-and-rescue effort began, with more than sixty people searching daily using helicopters, planes, and dogs.

Initially, rescue parties scoured trails near the campground, south of Joshua Tree, but Rosenthal's tracks were spotted on Tuesday, leading toward the canyon where he was found, some eight miles off course. On Thursday, six days after he'd set out, the helicopter located him.

Rosenthal had lost twenty pounds and was severely dehydrated; he vomited all over the helicopter when given water. At the hospital, doctors discovered his kidneys were in grave danger of failure, and he spent days in the intensive care unit. He sustained permanent heart damage but otherwise rebounded well. Within months, he was back hiking.

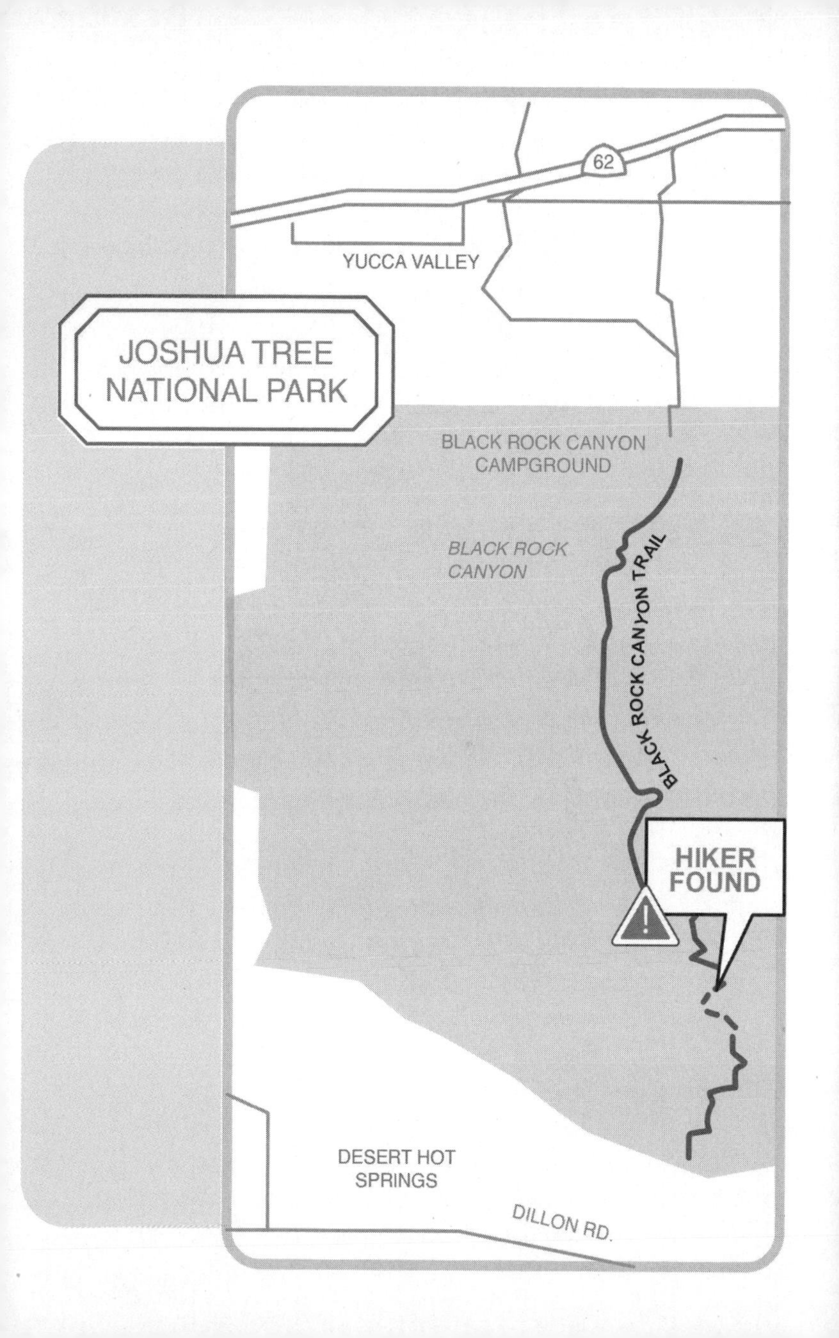
62
YUCCA VALLEY
JOSHUA TREE
NATIONAL PARK
BLACK ROCK CANYON
CAMPGROUND
BLACK ROCK
CANYON
BLACK ROCK CANYON TRAIL
HIKER
FOUND
DESERT HOT
SPRINGS
DILLON RD.

"Rehydration can be as lethal as dehydration. Introduce too much water too quickly, and your body won't absorb it properly. "If you're severely dehydrated, as Ed was, the gastrointestinal tract is one of the first things to go into stasis, because the body doesn't think it requires it," says Cummins. "If you then chug a jug of water, you'll vomit; your system isn't prepared for it." The best course of rehydration is intravenously, says Cummins, because you can correct electrolyte disturbances in a controlled manner. "The interplay of electrolytes and free water [water without sodium] is very complex. During rehydration, you need to watch your sodium levels closely," he says. Too much sodium relative to your water levels can lead to a condition called hypernatremia. Left untreated or if treated too quickly, this can cause your brain and lungs to swell dangerously, which can be fatal.

HOW TO SURVIVE BEING LOST IN THE DESERT

BE PREPARED. Rosenthal's backpack of gear helped ensure a greater chance of survival. Pack all the basics, including a first-aid kit, flares, windproof fire starters, a knife, water-purification tablets, and a headlamp. Always toss in extra layers, as well as two plastic tarps, one you can use to collect water and another for protective cover.

SHARE YOUR PLANS. Assume your cell phone won't work in the desert, and let a few folks know where you're headed and how long you expect to be gone. Give them a check-in time—if they don't hear from you by that time, they should alert the authorities. Leave a note with your planned route on the dashboard of your car too.

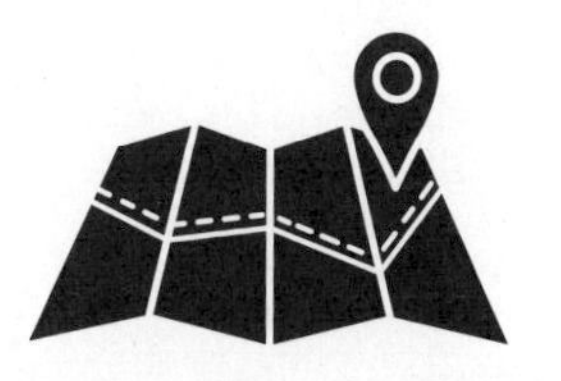

BRING A LOT OF WATER. In one-hundred-degree heat like Rosenthal braved, the body can lose up to two liters of water per hour. Drink water often and make sure you have at least one liter per hour of hiking.

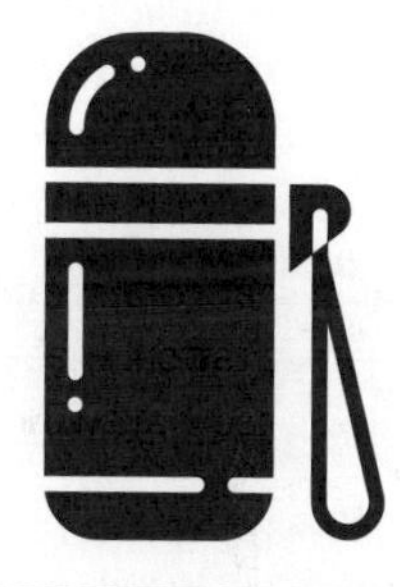

AVOID THE SUN. Easier said than done if you find yourself truly in the middle of a barren sandy desert, but any rocks and trees will offer some shade. Always wear a hat, as Rosenthal did, and resist the temptation to remove layers if you start to overheat; you need to protect your skin from the sun.

EAT SPARINGLY. The more you eat, the more your body must work to digest the food, a process that requires a fair amount of water. Eat slowly, and over longer periods, to keep your body from getting too dehydrated.

IF YOU NEED TO MOVE, MOVE AT NIGHT. The daytime heat of the desert can quickly dehydrate you and sap your energy. So if you have enough nighttime visibility to do so safely—ideally, your own light source, though the moon can also be bright enough at times—limit most of your movements to after dark. It'll be colder, and you may be tired, but the conditions are better without the sun battering you. Better yet . . .

STAY PUT. Your best bet for rescue is to remain in one spot. If you've shared your planned route and become lost, searchers will start with your original path of travel and spread out from there. If you keep wandering, as Rosenthal did, you could be miles away and it could take longer to be found.

FIND WATER. Rosenthal's attempt to drink liquid from cacti and yucca isn't recommended. Any water found in a cactus—and there's very little—isn't pure enough to drink. There are toxins that can cause more harm than good. Instead, look for areas of vegetation and try digging for water. Dig at least one foot down; if the hole is dry, move to another spot. If the hole is moist, dig deeper and widen the hole. After a few hours, water should appear at the bottom. Collect it with a cloth and then use water-purifying tablets before drinking it.

MAKE A SOLAR STILL. Another way to get water from the ground involves a plastic sheet. In a sunny spot (this won't work in the shade), dig deep enough to reach moist subsoil, then throw in nonpoisonous vegetation and plants. Place a cup in the middle of the vegetation, then place plastic on top of the hole. Seal the edges with loose dirt, and place a rock on the plastic, so the lowest point of the plastic is directly above your cup. The sun will draw out water from the vegetation. The water will condense on the plastic, then drip into the cup. Solar stills will also work with urine, per Cummins. "Ed could've put urine in a solar still and gotten potable water from the moisture evaporating," says Cummins.

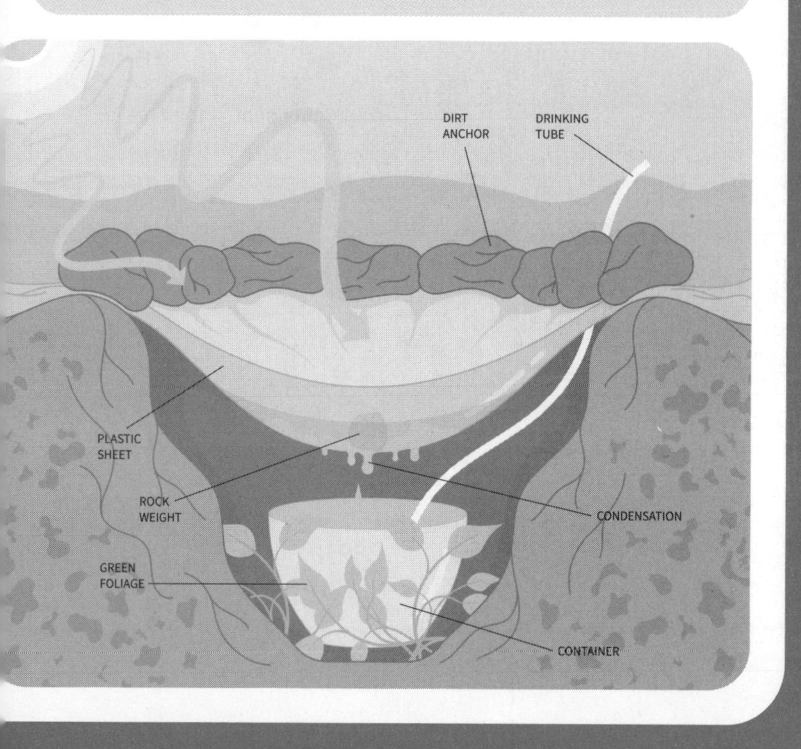

WHITEOUT IN THE MOUNTAINS

3

MAY 11. EVENING.

Ralph Summers scrapes at the ice above his head with a small metal shovel. He's exhausted but he can't stop now. For the past hour and a half, he's been digging a snow cave in a large snowdrift near the summit of Oregon's Mount Hood, which has been hit by a late-spring blizzard of such magnitude that this shelter may be the only thing keeping him and his fellow climbers from dying tonight.

Summers is thirty years old and makes his living escorting hiking parties up Mount Hood. At 11,249 feet, it's the state's tallest peak. Getting to the summit doesn't require a technical climb, but it can be tricky, especially if the weather changes quickly, like it did several hours ago.

His clients on this climb are a group of high-school students and adult chaperones from the Episcopal School in Portland, seventy-five miles west. They're part of a program called Basecamp, started by one of the school's teachers, Father Tom Goman, who believes that overcoming challenging environments—like summiting Mount Hood—fosters personal growth.

Over the course of the climb, with dark snow clouds bearing down on the peak above them and frigid winds picking up, several members of the group turned back. But Goman urged the rest of the group to keep climbing. They were at

10,700 feet, with just another 500 feet to the summit. Summers should have put up more of an argument and insisted they all turn back. They were more than 4,000 feet above the tree line, and Mount Hood's summit at this time of year was still mostly covered in snow. If the weather took a turn for the worse, they would be exposed on the mountainside, facing a difficult descent down steep slopes covered in ice and snow.

Eventually, the conditions forced them to turn back—but by then it was too late. The whiteout had utterly obscured their route, and they could no longer retrace their steps back down the mountain. Visibility had been reduced to just a few feet in every direction, and the trail wands they had used to mark their path had all been buried in drifting snow. They were, to all intents and purposes, lost.

Now Summers drops his shovel and swings his flashlight around the snow cave's interior. He's cleared out a space six feet by eight feet, and just high enough to sit inside upright. It's barely the size of a four-person tent. But the cave needs to fit thirteen people—ten students, Summers, Goman, and one other adult chaperone.

Outside the snow cave the wind is howling. Gusts now exceed 100 mph. It's close to eight o'clock, past dark. He can hear the eerie whistling through the entrance to the cave. He can't leave the others exposed any longer on the mountainside. Hypothermia is spreading now among the kids. He swings his flashlight around one more time, surveying the meager shelter he's managed to construct. This will have to do.

He crawls through the cave's entrance, a narrow tunnel about two feet long and three feet wide, purposefully constricted to protect against blowing snow and wind. He pops out, steadies himself against a gale that nearly topples him, and whips his light around. All Summers sees is white. His beam lands on the group, huddled under a tarp that's already covered in a foot of snow.

"Let's go," Summers shouts. "One at a time; students first. Hurry!"

Molly Schula is among the first. The athletic seventeen-

year-old senior has summited Mount Hood several times before, but she's weary and weak now. As is Giles Thompson, a burly sixteen-year-old sophomore, who follows. One by one, the trembling students crawl into the tiny space.

Outside, over the howling wind, Summers yells to Goman, "Is that everyone?"

"Pat's under the tarp," Goman replies. "Can you grab him?"

Summers trudges over to the tarp. He heaves it up and sees Pat McGinness in his sleeping bag. Two hours ago, the fifteen-year-old track star's speech slurred and he collapsed, no doubt suffering from some combination of hypothermia and altitude sickness. Now Pat looks pale and lifeless, though he's conscious. Summers hooks his arms under Pat's shoulders, lifts him to his feet, and practically carries him to the cave entrance. They both crawl inside.

> "Hypothermia is an abnormally low body temperature, caused by prolonged exposure to cold. First your body shivers, trying to generate heat, and your consciousness becomes impaired. You feel exhausted and confused. Experts later concluded that Father Tom Goman was likely suffering from hypothermia, which may have contributed to his decision to not turn back sooner in the face of the oncoming storm. As it advances, hypothermia leads to slurred speech and drowsiness, as Pat McGinness experienced."

"Altitude sickness happens when climbers ascend a mountain too rapidly. Your body doesn't have enough adjustment time to cope with reduced oxygen and the changing air pressure. Acute mountain sickness, the mildest form of altitude sickness, happens around nine thousand feet, per Deepak Sachdeva, an emergency room physician. You'll feel a headache and fatigue. "It'll feel like a hangover at first," says Sachdeva. High-altitude pulmonary edema is the next phase, where you start coughing, and fluid can build in your lungs. The last form, high-altitude cerebral edema, happens above fourteen thousand feet. "Your brain starts swelling, causing confusion, seizures, and a coma," says Sachdeva, who adds that these aren't sequential; you can experience them in any order.

Summers closes his eyes. With thirteen bodies mashed together, it's claustrophobic. The occupants gulp for air. Limbs jab other limbs. A panic is settling over the group. They've been on the mountain now for eighteen hours, since setting out at two thirty that morning for the climb, hoping to stay ahead of the weather. They're exhausted and frightened for their lives.

"If you're struggling to breathe, panic can set in and you hyperventilate. When panicking, your body switches from parasympathetic ("rest and digest") to sympathetic mode ("fight or flight"). "Slow deep breaths can help reverse this," says Anthony Giovanone, a doctor of osteopathy and a psychiatrist. "The base of your brain is reptilian, and it controls automatic functions like breathing and heart rate," says Giovanone. "Breathing is automatic, but you can take control of your breath, and if you can breathe slower, with a longer exhale than inhale, it tells that portion of the brain to slow down, and moves you back into parasympathetic mode, which helps settle the panic. Then your prefrontal cortex can reengage and increase the brain's ability for more complex thinking to handle the situation at hand."

Worse, Summers's limbs are already sluggish. He goes to clear icicles from his mustache, but his hand is slow in responding. Not a good sign.

Lethargy is an early sign of hypothermia. It can manifest as sluggish limbs and general drowsiness, but it's happening from a decrease in blood flow, per Sachdeva. As it worsens, your breathing becomes shallow. "Your brain is shutting down, and it can't regulate your lungs. You won't be getting enough oxygen to your tissue, and in a matter of minutes, that tissue will start to die," Sachdeva says.

* * *

By the yellow glow of her flashlight, Molly Schula can see drops of water falling from the ceiling of the snow cave. They've been in here for a few hours, and their combined breathing in such a confined space is melting the walls. An icy puddle of slush spreads on the floor. On her back, the pool of ice water is soaking through her jacket. Rolling a bit, she sees her classmate Patrick McGinness. His hypothermia seems to be easing; he's stopped shivering. *Small victory there*, she thinks.

> **Children and teenagers are affected by hypothermia faster than adults. "The younger you are, the greater your total body surface area [relative to your mass], so you lose heat faster," Sachdeva explains. In severe cases, as the climbers experienced inside the snow cave, hypothermia causes your pulse to weaken and your breathing to become labored. "Your body is trying to conserve energy," says Sachdeva. "Your heart will slow, which means less blood reaches the brain, so you're less alert." Your body shivers to create heat to warm you, but as the brain shuts down, it can't create that compensatory response, so you stop shivering. "That's actually a bad sign," says Sachdeva.**

It's too cramped, so they decide to take turns standing outside in the blizzard. Three people at a time, in twenty-minute shifts. Finally, Molly has room to sit up and breathe. She feels her heart rate slow, her breathing calm. But where is Father Goman? He went outside for his shift almost an hour ago.

Finally, she sees him appear in the entrance tunnel. He's in horrible condition. His hat has blown away, and his head is covered in ice. Lips blue and face pale, he shudders violently. He needs to warm up. Fast.

> When experiencing hypothermia, quick actions are needed. Remove any wet clothing, and warm the center of the body first—chest, neck, head, and groin. If no warm clothes are available, skin-to-skin contact works. Warm drinks should be given only if the person is responsive and able to swallow. In severe cases, where the victim is unconscious, handle them gently. "The heart can get into an irritable state, where sudden jerky motions can send the heart into fibrillation, where it's not beating regularly, then you're not pumping blood," says Sachdeva. He suggests avoiding dragging someone down a rocky slope. "Spend a minute checking for a pulse in the neck; if you don't feel one, perform CPR if you have proper CPR training," he says.

Bodies shift to allow Goman to lie down out of the ice pool. Molly borrows a hat from a classmate and pulls it over her teacher's head. Ralph Summers drapes a sleeping bag over him, and Molly tucks it around him. Goman stammers thanks, saying he feels better now.

But his face is turning bluer by the second and his shivering is increasing. He's far from okay.

As the night wears on, the students take turns using the group's only shovel to keep snow from accumulating over the cave's entrance. One of them is outside digging when Molly hears him cry out. A minute later, he crawls back inside the cave, a look of panic on his face. "The wind ripped the shovel out of my hands," he says. "It's gone."

Molly and the others process this. If the entrance ices over completely, their shelter will be their tomb.

* * *

A ringing phone at one thirty in the morning wakes Rick Harder at his home. The thirty-four-year-old is a paramedic for the Portland Fire Department. He's also on call for mountain rescues, thanks to his experience as a pararescue jumper in the air force. And his skills are badly needed now. A group of high-school students is stranded near the top of Mount Hood.

Harder gathers his gear and heaviest parka. The storm isn't expected to lift for another day and he'll need extra protection. He makes it to Timberline Lodge, halfway up Mount Hood, where the search-and-rescue base camp is being staged, around four in the morning. The climbing party was last seen ten hours ago by the climbers who turned back. Harder grills them now: Where were they specifically on the mountain? What gear did they bring? How is everyone dressed? The answers are vague and unhelpful.

The team's leader says that no rescue attempt will start before dawn. The weather is just too severe. Harder doesn't like this call—teens lose body heat faster than adults, so advanced hypothermia comes quicker—but he understands. Conditions are terrible, so leaders hope daylight will provide at least some visibility. The triangular search zone is two square miles, so the crew splits into four teams. Harder will lead Team One.

They start up the mountain a few hours later, at seven o'clock in the morning. Blasts of wind, ice, and snow force Harder and his seasoned crew to their knees at points. It's slow going, trudging through thick snow, each man laden with heavy bags of medical supplies, but two hours later they hit 9,700 feet. They need to press on, higher, and more to the east to cover their section of the search zone. However, in these conditions, advancing is treacherous. Visibility is zero and it's bone-chillingly cold. Snowcats, fifteen-hundred-pound enclosed vehicles with tracks instead of wheels, are flipping over in the high winds.

He grabs his walkie-talkie, crusted in ice, and radios base camp at the Lodge. "This is Harder. Don't bring any teams up here! It's too dangerous."

Harder and Team One struggle for another three hours. The team is freezing, and there's nowhere to stop and add extra layers. When the wind blows one man to the ground, stripping a glove in the process, Harder knows it's too brutal to continue. If the storm intensifies, the rescuers may need rescuing. He orders the team to return to base.

The kids will be on their own for countless more hours. And if Harder and his seasoned crew can't brave this storm, what hope is there for the group trapped in it?

* * *

At seven in the morning, Ralph Summers peers out of the cave's entrance, hoping to see daylight. There is none. The storm is worsening. He looks at the group, damp from sitting in the ice pool for the last twelve hours. It must be just barely above freezing in here. Everyone's teeth chatter, and many shake uncontrollably. Few have eaten. Goman's quivering blue lips are covered in frost. Summers asked him to count to five, but he couldn't. Hypothermia and lethargy will kill Goman soon—then everyone else unless he acts. Now.

"I'm going for help," Summers says. "I need someone to come with me. It'll be safer with two of us."

Silence. Summers doesn't want to go alone, but he may have no choice.

"I'll come, Ralph." A shaky voice from the back of the cave. It's Molly Schula.

Summers smiles at the teen's bravery as the two squeeze out of the cave. Guided by Summers's compass, they trudge down the mountain. Some slopes are too steep and icy to stand upright, so the two slide down, at sometimes terrifying speeds. Molly screams as they slide. The teen's hair is crusted with ice, her steps are sluggish, and she's weak—she's succumbing to exhaustion, Summers realizes. Walking will keep her alive, so Summers keeps pushing her.

Around nine o'clock, they spot a tree. Summers knows that this means they've reached six thousand feet—the tree line. That means they're descending, so Summers excitedly concludes they must be on the right path. They still need to

cross a canyon, but help is within reach. They're aiming for the Timberline Lodge.

"Keep walking, Molly," Summers says. "We're almost home free."

* * *

Giles Thompson studies the snow cave's opening. It's constricting by the minute, now about the width of a car tire.

With their shovel lost in the storm, the only tool they have is Summers's ice axe. Giles grabs it and starts to chip away at the icy walls.

Three students need to relieve themselves. They tell Giles they'll try to widen the opening to the tunnel as they crawl through it. After they've left, Giles lies on his back, arms wedged against his sides, and scrapes at the tunnel's ceiling and walls. Icy shards cover his face, but he moves as quickly as he can. It's hard to tell if he's making progress.

Fifteen minutes pass. A shout from outside signals that his three classmates are coming back. Thompson looks at the entrance at the far end of the tunnel.

What he sees horrifies him. The outside end of the tunnel is nearly closed, covered in built-up snow that's solidifying to ice.

A boot punches through. Giles grabs it and pulls hard. It comes off in his hand as he falls back on the cave's floor.

He hears a voice yelling from outside: "No! Give my boot back!"

But, like an iris closing, the small opening seals, and Giles, weak and dehydrated, can't reopen it. He hears the three students' bloodcurdling screams over the roar of the wind.

They're trapped out there. And now Giles and seven others are stuck in here.

* * *

Ralph Summers squints. There's something dark overhead, in the distance. It's a chairlift.

"We're saved!" Molly Schula exclaims weakly.

"We can't celebrate until the whole party is off the mountain," Summers says as they reach the chairlift. He brushes away snow covering a sign on the lift's support column. Summers is dismayed. It's not the lift he'd thought; they're two miles off course from the path down to Timberline Lodge.

"How much longer, Ralph?" Molly asks.

"This still leads to civilization," Summers says. "Another thirty minutes. You can do it."

But he's not sure if she can. She's staggering, her limbs barely cooperating. Summers wraps an arm around the teen to steady her as they move. At last, a building emerges through the whiteout. The ski school office. Just after ten o'clock they burst in, exhausted, windburned, coated in ice. Both survivors can barely stand. Summers identifies himself to the workers and within minutes, it seems, rescuers arrive and pepper him with questions.

Summers's vague answers about the snow cave's location worry rescuers. No one in the party used an altimeter, so pinpointing the exact altitude to search is out. The best chance is to have Summers try to locate the trapped group from the air.

The helicopter takes off around one o'clock. The storm is still going. Summers hasn't slept for thirty-six hours. He slaps his face to stay alert. The rescuers in the chopper are relying on him, but he's struggling. In the six hours since he and Molly left the cave, two feet of snow have fallen. The terrain is unrecognizable—just blankets of white. The chopper circles for hours, until another whiteout arrives, forcing it to land.

Over the helicopter's headset, a rescuer says what Summers is thinking: "Those kids have to spend another night on this mountain."

* * *

Rick Harder's leg nervously bounces in the back of the helicopter. It's six o'clock the following morning. The storm, which stymied search efforts throughout the night, has finally lifted, bringing clear visibility. And a team on the mountain has finally spotted what might be a sign of the students—a red sleeping bag, near a canyon rim.

The helicopter carrying Harder and Ralph Summers is racing to the area. Harder sees a dot of red as they land. He can already tell it's not a sleeping bag.

It's the body of a male student, frozen in the fetal position. He's missing a glove and one boot. Yards away, they find another body. Harder's radio chirps; it's another helicopter, farther up the canyon. They've spotted another student in the snow, but from the air they can see no sign of movement. Harder sighs. *That's three. But eight could still be alive.*

For the rest of the day, Harder and Summers fly around,

searching for the snow cave. It's close to four o'clock when Summers perks up. "Hey! I recognize that ridgeline. Can we land?"

The helicopter's low on fuel, so the pilot asks if Summers is sure. Watching him hesitate, Harder says, "You only have to be right once, Ralph." Summers nods, and the chopper's skids hit the snow.

Summers seems positive this is it, so Harder calls for backup. More rescuers arrive, by helicopter and on foot,

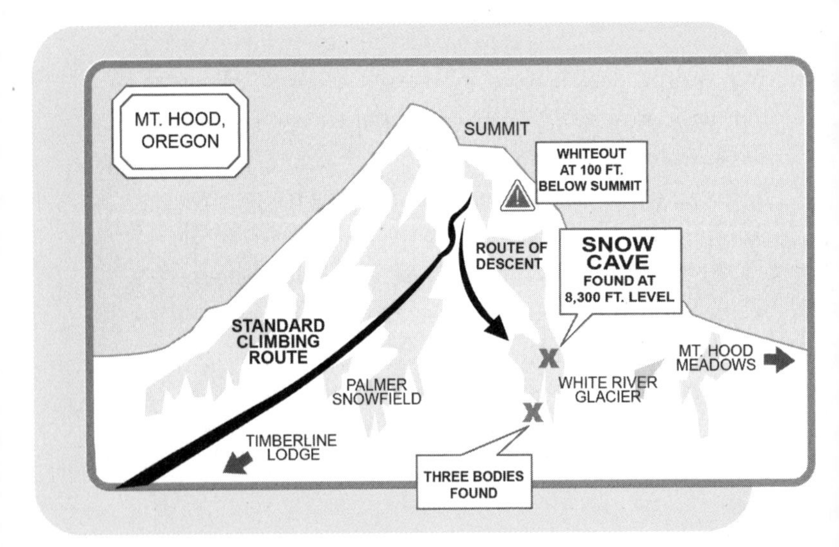

forming a line and probing the snow with long aluminum poles as they march across the mountainside. It's tedious, but after an hour, one probe hits something soft.

"I need a shovel," shouts the rescuer. "Now!"

Rescuers swarm the area, digging madly. Four feet down, the shovel breaks through to an open space. A horrible stench rises from the dark hole as Harder and crew peer down.

They hear a faint moan. Harder locks eyes with Summers. There are survivors.

But only two.

* * *

The 1986 Mount Hood tragedy claimed nine lives, making it the second worst alpine disaster on record in North America. Father Tom Goman and Patrick McGinness were among those who died on the mountain. Giles Thompson survived, though his heart had to be massaged to get it beating again. Due to frostbite, doctors amputated both of his legs—one below the knee, one above it. Only one other student made it out of the cave alive. Molly Schula escaped serious injury too.

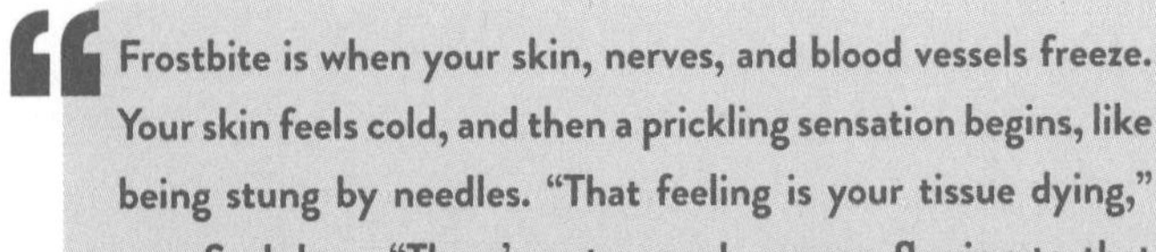

Frostbite is when your skin, nerves, and blood vessels freeze. Your skin feels cold, and then a prickling sensation begins, like being stung by needles. “That feeling is your tissue dying,” says Sachdeva. “There’s not enough oxygen flowing to that area.” As frostbite progresses, the skin goes numb and then gets inflamed and discolored. Your skin turns white or blue during severe frostbite, and you lose all sensation in that area. “You have minutes to hours to reverse this,” says Sachdeva. If left untreated, that tissue turns black and hardens as it dies.

Ralph Summers was haunted by the death toll but praised for digging the cave and saving the lives of three of the teens. Rick Harder spent another decade as a search and rescuer, saving more than three hundred people during his career.

HOW TO SURVIVE A HIGH-ALTITUDE WHITEOUT

Whiteouts are overwhelming because you can't tell sky from ground. This disorientation alone is enough to cause dizziness and nausea. Add in hypothermia-induced confusion and the symptoms of altitude sickness, such as nausea and fatigue, and it's a deadly mix. Here's how to escape a whiteout.

BRING PLENTY OF GEAR. Extra layers, sleeping bags, and a cooking stove are all vital for staying dry and warm. Make sure your bag is weather- and waterproof; you want these items to remain dry. Bring a metal shovel in case you need to dig a shelter.

USE GPS TO PLAN AND SHARE YOUR ROUTE. In whiteout conditions, you may be led off course, but if you've got a GPS—and shared your route ahead of time—searchers will have a better idea of where to look for you. Make sure your GPS has an altimeter, so you can relay that information if you're stuck.

DESCEND TO COMBAT ALTITUDE SICKNESS. The mildest form, acute mountain sickness, can clear up with ibuprofen, says Sachdeva, but for serious altitude sickness, you should get to a lower altitude to reduce risk.

IF THE STORM DOESN'T PASS, STAY PUT AND DIG IN. Don't try to feel your way along a route. Falling snow can mask deep crevasses or conceal steep drop-offs. And the lack of visibility can lead you to wander in circles. Seek shelter on a leeward side of the mountain—the side protected from the wind—and stay there. (See below for how to create a shelter.)

DESCEND ONLY IF YOU'RE CONFIDENT. Whiteouts are often limited to certain altitudes, so descending the mountain can help you escape one. However, you must be certain you're going in the right direction. If you can follow your own steps back down, do so. Trail wands—colored sticks placed in the snow at regular intervals—can help you retrace your path. The Oregon Episcopal School group used these, but the storm was so severe the wands were buried before they descended. If you can't determine the right direction, throw a snowball. If it seems like it lands in midair, that slope is rising. If the snowball lands below your feet, that's downhill. If the snowball vanishes, be extra careful; it's likely there's a deadly drop near you.

DIG A SNOW CAVE. If you need to make your shelter, do so in an area free of rocks and away from potential avalanches. Find a snowdrift that's at least five feet deep. Make sure the snow isn't light and powdery—that will collapse on you. You want it to be a little hard. (If you can't find good snow, consider digging a trench and covering yourself with a tarp.) While you dig, leave one person outside with a shovel. In case of collapse, they'll be able to save you. Dig at least ten feet in diameter to fit three or four people.

Make sure to dome the ceiling, to avoid water dripping on you inside. Leave at least one foot of snow at the top, to lessen the risk of collapse. Dig the entrance tunnel to the cave at least three feet wide and downhill, to stop blowing snow and wind. Make the floor of the cave higher than the entrance; it'll stay warmer that way. Last, you can add a ventilation hole to help with breathing—but poke the sides of the cave, not the ceiling.

STAY WARM AND DRY. To fend off hypothermia, ditch wet layers and put on dry clothes. Warm vital parts of your body first: the head, chest, neck, underarms, and groin. Bring layers made of wool or wool synthetics; they will be warmer. Cotton can trap moisture and exacerbate feeling cold. Loosely add layers because warm air gets trapped in between. Tight clothes make you feel colder. Warming up dying tissue from frostbite will be extremely painful, warns Sachdeva, who adds you're unlikely to get tissue warm enough in the field to feel this level of pain.

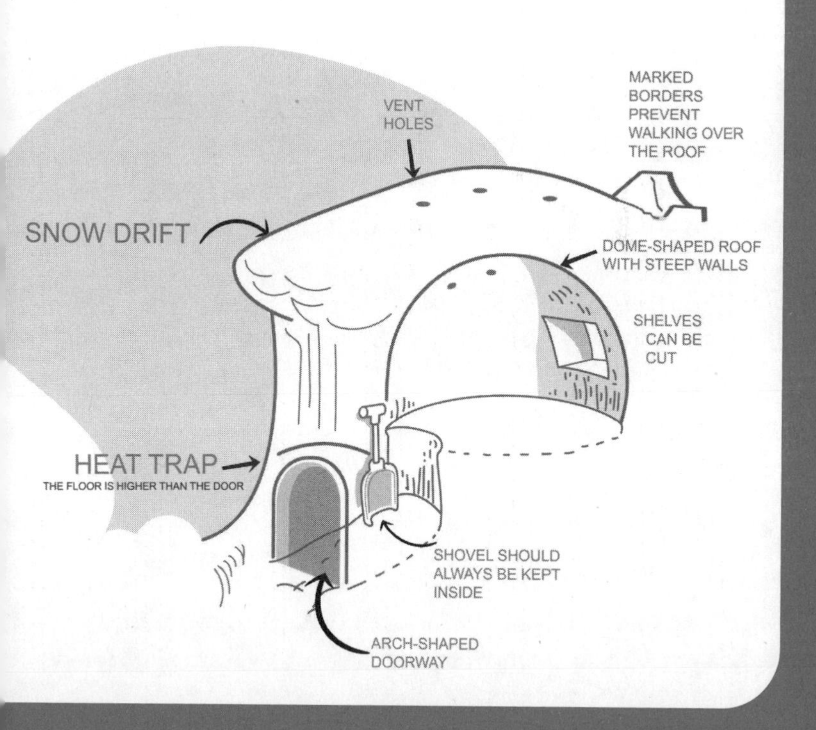

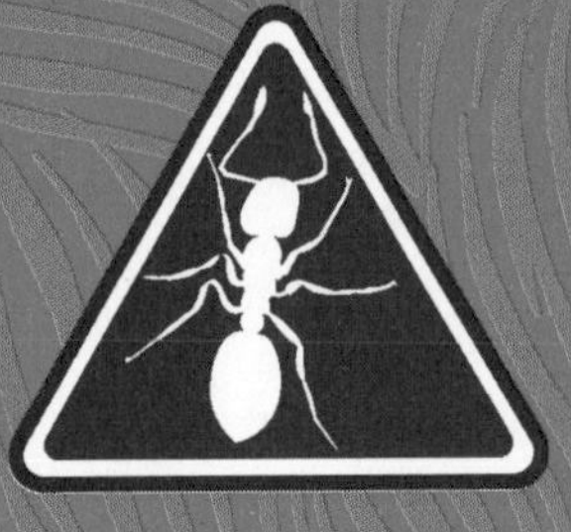

LOST IN THE JUNGLE

4

DECEMBER 1. MORNING.

"Go right! We have to go *right*!"

Kevin Gale's shouts are barely audible above the sound of the rapids on the Tuichi River, in the jungles of northern Bolivia. Yossi Ghinsberg hears them and dutifully plunges a handmade oar into the raging water. But the rapids are too strong; the oar does nothing against the current. Their eight-foot raft, made of balsa logs the two men lashed together, is headed directly toward an enormous boulder.

Ghinsberg drops flat, grabbing the leather straps holding the raft together, and braces. Water washes over him as the raft slams into the boulder and jerks to a halt. The current has pinned the raft to the boulder, and now they're stuck. Ghinsberg glances at Gale, who's clinging desperately to his side of the raft. Ghinsberg looks left and sees something that makes his heart race: a twelve-foot waterfall, just downriver.

Even if they can shake the raft free of the boulder, it'll be swept over the side—with them on it.

"I'm going to swim to the bank," Gale yells over the noise of the rapids. "Then I'll cut a vine down and throw it to you, then pull you to shore." Ghinsberg gulps and nods. The twenty-three-year-old Israeli has zero outdoor experience. Fortunately, Gale, a twenty-nine-year-old adventurer from the Pacific Northwest, is beyond experienced in the wild, and Ghinsberg trusts him completely.

Clinging to the raft, Ghinsberg watches Gale swim to shore, fighting the current all the way. On the shore, Gale uses a machete that he'd thrown to the shore from the raft. *Thank God*, Ghinsberg thinks. *This'll be over soon.* But before Gale can throw him the lifeline, Ghinsberg feels the raft shifting beneath him. It's wobbling, threatening to break free.

"Hang on, Yossi!" Gale shouts from shore. "You're going over! Don't let go!"

Suddenly, the raft slides from the rockface and is plunging down the waterfall. Ghinsberg tightens his hold on the raft's straps and squeezes his eyes shut. Then he's submerged, the noise of the rapids replaced by gurgling as he prays for air. His prayers are answered when the raft bobs to the surface. It's some kind of miracle: Ghinsberg is uninjured. He whips his head around and sees Gale running down the riverbank, shouting at him as he's swept downriver. Just seconds later, Gale has receded from view.

No, no, no. Ghinsberg thinks. *How the hell am I going to survive out here alone?*

* * *

Yossi Ghinsberg met Kevin Gale by chance during Ghinsberg's trip through South America in 1981. They befriended two other backpackers, and the foursome embarked on an expedition to hunt for gold and precious metals in the jungles of northern Bolivia, part of the Amazon rainforest basin.

But with rations running low, two of the men turned back just this morning. Gale persuaded Ghinsberg to continue as a duo. His idea was to float the raft one hundred miles down the Tuichi River, to a town called Rurrenabaque, then catch a flight back to La Paz, Bolivia's capital. A lifepack lashed to the raft contained all their survival gear, including a first-aid kit, map, mosquito nets, flashlight, ponchos, lighters, and a limited amount of rice and beans. Now, just hours since they set off down the Tuichi, Gale's plan has already led them to disaster.

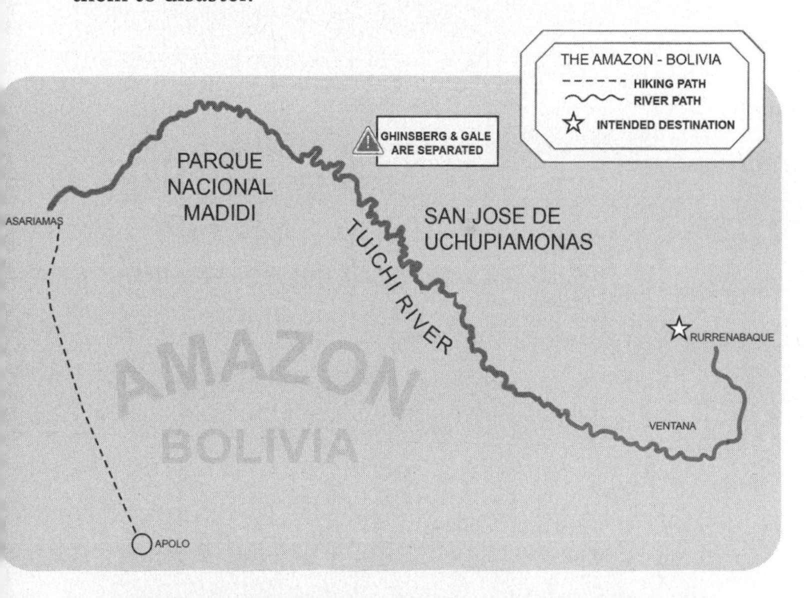

As Ghinsberg floats alone down the river, he's relieved to see that the lifepack is still tethered to the raft. Then he hears the unmistakable roar of another waterfall approaching. He can't gauge the drop distance, but judging from the sound, it's a big one. He wraps his fingers around the straps. A second passes. Then there's a terrible bang, and he's flung off the raft, into the air. He waits for the sickening crunch when his body hits the rocks below. But once again his luck holds and he splashes down in the water unscathed.

Ghinsberg climbs back on the raft just in time to see the lifepack float away. It must have been knocked loose going over the waterfall. He stares helplessly as the current whisks

him away from everything he needs for survival. He needs to go back for the bag, but he can't get off the raft here; the rapids are still too swift, too dangerous. He waits until the river calms, then abandons the raft—it's too heavy to drag ashore—and wades to the bank. He doesn't want to give up the lifepack; without it, he'll surely die.

He clambers up a canyon wall to get a better vantage point. From here, he spots it: the lifepack, bobbing in an eddy where it's washed up against some rocks, maybe three hundred yards upriver. But the route along the skinny shoreline is too slippery and dangerous to attempt now. It's getting dark. And he's exhausted. His best bet, he decides, is to find shelter for the night and try for the pack in the morning.

* * *

Ghinsberg finds a small recess in the canyon wall and gathers branches to pile near the opening. He doesn't want an encounter with a jaguar; Gale had mentioned they prowl the area. While he arranges this ramshackle barrier from inside, the gravity of his situation washes over him. No food. No gear. No survival skills. He has only the soaked clothes he's wearing. He sobs as he tries to sleep, rocks poking his back, panic building in his mind as the loud sounds of the jungle's night—howling monkeys, screeching birds, buzzing insects—fill his ears. There's little rest.

The following day, he manages to make his way back upriver and retrieve the lifepack. He celebrates with a meal of rice and beans. Over the next few days, he keeps going upriver, hoping to locate Gale. But soon his food runs out, and meals become meager: a few cloves of garlic, a lemon, and some foraged jungle fruit.

After days of trudging through the jungle along the river, Ghinsberg's feet are killing him. Sheltering in a cave one evening, he tugs off his soaked hiking boots. Red and yellow stains cover his socks, and the stench is horrendous. He peels them back, agonizingly. Pus and blood ooze from his toes.

"Trench foot occurs when your feet are exposed to moisture for extended periods. "Skin is designed to be dry," says Matt Cummins, an emergency physician. "When it's wet, it starts to break down and cause issues." First, your feet itch and tingle, then the ailment progresses to numbness. Feet may turn bright red and feel like heavy blocks of wood. You'll have difficulty walking. Eventually, they'll become pale and clammy, as the blood vessels in your feet constrict. It can happen in as little as ten hours, though it usually takes days.

As the foot tissue breaks down, it'll lead to bleeding open wounds, and then infection, as Ghinsberg experienced. To avoid this, "dry your feet as fast as you can as often as you can. Put them near a fire, if you can, but keep them at a safe distance. Your feet will be numb so you may not feel any burning if they're too close. If they're getting red, apply antibiotics," says Cummins.

There's no Vaseline in the lifepack, but there is a cream—insect repellent. It'll have to do, so he slathers it on. The pain is intense and this, along with a fever, keeps him up at night.

> **Was Ghinsberg's application of insect-repellent cream a wise move? "It depends on the base of the insect repellent," says Cummins. "If it's alcohol based, that's bad. It will dry the skin and cause more problems, like your skin cracking. If it's oil based, like an ointment, it can create a water barrier and can be helpful. However, nothing beats getting your feet dry."**

* * *

Ghinsberg rests for two days in the cave, letting his feet heal and regaining his strength. When he's ready to move again, it's December 6, his sixth day alone in the jungle. He abandons the idea of tracing the banks of the Tuichi River, searching for Gale, in favor of taking what he thinks is a shortcut over a mountain ridge a few miles away. But he gets lost in the hours-long hike to the mountains, and when he fails to track the direction of the setting sun, he has no idea which way to go to reach the river again.

A mosquito net tied to tree stumps is his tent for the night, and he beds down on some gathered leaves and uses another mosquito net as a blanket, trying to block out the sounds of the jungle. Just before he drifts off to sleep, a nearby rustling spooks him. Ghinsberg bangs a metal can and shouts, but the noise grows closer. He snaps on a flashlight and casts the beam around. It falls on two glowing green eyes, twelve feet away. A jaguar. Its black spots and gold fur shine under his beam, and its tail swishes slowly.

Ghinsberg shouts at the animal, but it just stares in his direction. So he grabs a lighter and aerosol can of bug spray and creates a flamethrower. A long stream of fire shoots at the jaguar. It singes the hair on Ghinsberg's hands but he doesn't stop until the can is empty. It does the trick; the jaguar's gone. He switches off the flashlight with shaking hands. But he won't sleep tonight.

* * *

By his tenth day alone, Ghinsberg is starving. The rice and beans turned moldy a few days ago, so his only food comes from scavenging small amounts of fruit and plants. He forgets his growling stomach when he happens upon a clearing and sees four thatched-roof huts. This must be Curiplaya, an abandoned gold-mining camp he'd heard about when planning the expedition. And if he's right about this being Curiplaya, that means a small jungle village called San José is only a few days farther. If he can reach it, Ghinsberg can organize a search party for Gale. But first, he needs sleep. He enters a hut and finds a wooden platform, raised off the floor. A bed! He falls onto it, exhausted, and sleeps. He spends two nights here, then decides to keep moving, to try to reach San José, where he can get a search party organized to find Gale.

> **Without food, your body becomes malnourished. This can lead to low energy, reduced muscle strength, poor concentration, and difficulty keeping warm. As it increases, your risk of infection rises, and your ability to recover from infections decreases. Wounds are slower to heal, as Ghinsberg would later discover with his spreading rash.**

Resuming his hike, Ghinsberg gets hopelessly lost for one day, moving in a giant circle. Then, just before the sky opens up with rain, he finds the Tuichi River again by pure luck. He also finds the remains of an old campsite on a riverbank—two poles tied with vines. He hurries to gather palm fronds to create a roof for this shelter. The rain is so torrential that it's uprooting enormous trees. Ghinsberg tries in vain to sleep, hoping a tree doesn't crush him, as the jungle seems to collapse around him.

In the morning, he's still uncrushed, but the rain continues to fall. Ghinsberg steps outside of his makeshift shelter and is aghast. The Tuichi is overflowing its banks and has widened by hundreds of yards. So much rain has fallen that a new river has formed nearby, streaming down the hillside into the Tuichi. It's too risky for him to cross no matter which way he wants to head. He huddles back in his campsite to wait out the storm. Pain erupts as he sits; the day before he slipped trying to scramble down a hill and impaled his buttocks on a dead tree branch. His tattered jeans are soaked with blood. He shifts uncomfortably, trying to take pressure off the wound.

> Open wounds like this can be extremely dangerous due to the high risk of infection. "Damaged tissue breaks down at various speeds, impacted by whether infection takes hold," says Cummins. "If infected, the decay is much faster. When a wound is caused by a nonsterile object, like a tree branch, it's very hard to keep that clean." Wading waist-deep in jungle rivers would make infection inevitable, says Cummins. "Even with bandages, it won't be waterproof or airtight; you'll have issues. When the skin dies, it'll increase the size of the wound, and you'll have pus and drainage, and nasty smells. If that infection spreads to the bloodstream, you'll become septic. That can kill you."

But the ground he's sitting on is getting wetter. He lies down, but then his back is soaked. He sits up and looks around. The water pooling around him is rising. The Tuichi and the new river are melding into one, creating a flash flood—right where he's sitting. He stands, but there's nowhere to go. The water's up to his knees. Then his waist. Then his chest. And the current is deadly strong. Ghinsberg grabs his lifepack and wades out from under his shelter. He finds a tree and wraps his arms around the trunk, clinging to it for dear life as the water continues to rise.

Finally, the flood subsides. His sodden boots squelch in the muddy earth. Ghinsberg has to keep moving, to get away from flooded areas. Stones and mud inside his boots rub against his decaying feet. Each step is maddeningly painful. It takes hours, but he finally makes it to a high bluff overlooking the Tuichi. The rain stops, but the river's still coursing strong.

* * *

It's been eighteen days since he lost Gale. Ghinsberg's body is rotting, he thinks. Red patches are all over his legs, visible through tears in his jeans. The rash on his legs that he'd first noticed a few days back is spreading. A cut on his wrist oozes pus. The wound on his backside still throbs. As do his swollen feet. When he'd last managed to tug off his boots, all he saw was a mess of bloody flesh. He's exhausted, and beyond hungry. When he'd caught a glimpse of himself in the water, he was shocked by his gaunt face. He must have lost thirty pounds by now.

Ghinsberg has given up on trying to reach San José. Instead, he thinks backtracking to the campsite where the flood started is the best place to wait for help to find him. The riverbank there is wide, and if it's not still flooded after the storm, he can be seen from the air. That's assuming there's anyone searching for him.

He trudges back, slowly. The rainforest floor is soaked and muddy, and when he steps into a puddle, his foot doesn't come back up. It's stuck. So is his other foot. Worse, they're both sinking into the muck. It's as if the rainforest is swallowing him whole. He's thigh-deep in seconds. Frantic, he tries grabbing on to reeds nearby, but they come off in his hands when he pulls. He's soon up to his stomach in mud.

The thought of drowning in mud horrifies him. Ghinsberg pushes his arms out, as if he were swimming, and tries to kick his legs. His feet move, ever so slightly. Again, he strains against the muddy earth. A little more movement. First it's millimeters, then inches. It takes all his remaining strength and more than an hour, but his fingers finally reach dry ground. He has to physically yank his legs from the muck, but he's free. His body is coated in thick black mud, but he's still alive.

* * *

Yossi Ghinsberg's struggle with the muddy bog leaves him so weak that he can't walk, so he crawls on all fours. He gathers palm fronds to make a bed for the night. He's so delirious from malnutrition that he hallucinates, imagining there's a beautiful young woman accompanying him in the jungle. When he realizes she's not real, he worries he's going insane.

> "A variety of things cause hallucinations, as Ghinsberg experienced—everything from a lack of food to insomnia to infections. "Malnutrition means Yossi's not getting enough sugar to his brain, so he's lacking fuel for the brain to properly work," says Anthony Giovanone, a doctor of osteopathy and a psychiatrist. "During infection, your body will have an inflammatory response, and in this state you can have neurodegeneration in the brain." Without adequate rest, toxic materials build in the brain. "As you sleep, the brain is bathed in waves—your neurons fire at certain frequencies and those firings are collectively called waves—which recycle toxic metabolites and reset neurons," explains Giovanone. "When this doesn't happen, the brain won't function properly." Each of these issues alone can cause hallucinations, but when facing all three, you're bound to perceive things that aren't there, says Giovanone.
>
> Hallucinations occur, to some extent, to ground yourself. "Yossi's not scared when he sees the woman in the jungle because it's comforting; it means he's not alone," says Giovanone. "His brain may have invented a maternal figure because, on a subconscious level, he's trying to envision a caregiver in the vein of his mother. When we're hurt or scared, on a basal level, we seek a parent's comfort.""

He's starting to drift off when he has an urge to urinate. He's too tired to move, so he just relieves himself in his jeans and falls asleep. Minutes later, there's a stinging pain in his thighs, like needles jabbing. In the darkness, his fingers find the culprit: an ant. As he plucks it off his skin, the pincers dig in. Hard. It feels like it's taking a tiny piece of flesh with it, but he finally gets the ant off. Another sting, on his waist. Another, on his knee. Another, on his scalp. Ghinsberg works hurriedly, flinging these minuscule assailants from his body. For each ant he's able to clear, four more seem to appear. Somehow, Ghinsberg falls asleep.

> **How could Ghinsberg fall asleep while being bitten by fire ants? "One factor is repetitive noxious stimulation. With repeated bites, your nervous system turns down the volume; it desensitizes you to each individual bite," says Beth Palmisano, a medical doctor who specializes in pain management. "Also, the body's adrenaline and endorphins kick in during these situations, and those endorphins are natural opioids, which dull pain. Natural endorphins can make you relaxed and sleepy too."**

When his eyes open to the gray dawn the next morning, he's stunned by what he sees—a rippling red wave on the ground around him. He leaps up. Fire ants, millions of them.

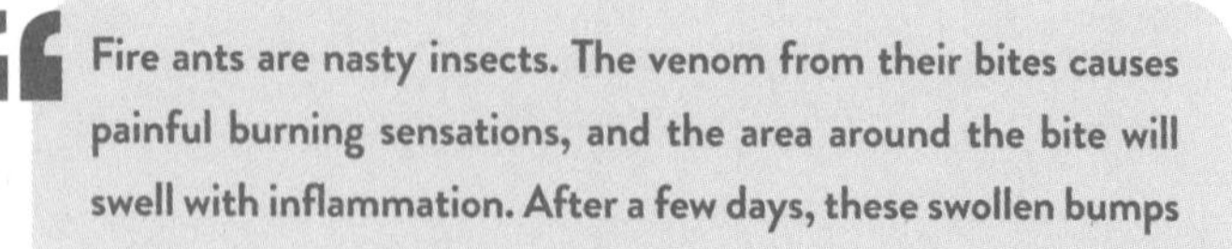

> **Fire ants are nasty insects. The venom from their bites causes painful burning sensations, and the area around the bite will swell with inflammation. After a few days, these swollen bumps**

turn into fluid-filled blisters. As they pop, they can become secondary sources of infection. "Blisters will itch as they heal, and the more you scratch, the more you break the skin open, which leads to an increased risk of infection," says Cummins.

In about 5 percent of fire ant attacks, the venom can cause an allergic reaction that leads to death. "Anything with a protein that your body isn't used to, such as ant venom, can cause a reaction," says Cummins. "How exaggerated your immune response is will vary from person to person; you could develop hives or experience full anaphylaxis, where your airways swell, and if they swell shut, you can die. It has little to do with the ant itself, though. It's more about your body's response."

They're everywhere, covering the ground, the mosquito net he uses as a blanket, and even up the tree trunks. Then he remembers something Gale said: that the smell of urine attracts insects. He did this to himself. He grabs his gear and limps away, spending fifteen minutes ridding his body and supplies of the ants.

He hobbles along the Tuichi River until he sees a small hut, which looks oddly familiar. Moving inside, he realizes he's been here before. It's Curiplaya, the abandoned mining camp. The flood must've washed away the other three huts, and this one's barely standing. But it's enough for Ghinsberg. He lies on the floor, wrapped up in a mosquito net, and tries to ignore the stabbing pains in his abdomen. His ribs are protruding and his guts feel like they're on fire; it must be malnutrition. His forehead throbs too. His fingers find a lump above his eyebrow. *Did I hit my head? I don't remember doing that.*

A buzzing sound interrupts his thoughts. Bees? He's probably imagining this too, like the woman. But the noise gets louder. He's not hallucinating. It's not the sound of bees; it's the sound of an engine! He flings off the net and stumbles to the river's edge. There, in the distance, he sees the source. It's a boat, with four people on board.

"Hello! Hello!" he shouts weakly.

The boat turns and heads in his direction.

"Don't move, Yossi! I'm coming!" A voice from the boat. Ghinsberg can't believe it. It's Kevin Gale.

As the boat nears the shore, Gale jumps into the water and rushes toward Ghinsberg, who collapses into Gale's arms.

* * *

After losing Yossi Ghinsberg over the waterfall, Kevin Gale hiked for five days and ultimately floated downriver on a thick log, until he came upon some fishermen. A few days later, he met with Bolivian navy representatives in Rurrenabaque to organize a search party for Ghinsberg. It took several attempts by air and water, but he finally located Ghinsberg. The rains that caused the flood on the Tuichi that nearly killed Ghinsberg were, ironically, the only thing that allowed Gale's boat to go this far up the river to find him.

Ghinsberg had wandered alone in the Bolivian jungle for three weeks. He'd lost fifty-five pounds and needed to have leeches removed from his skin.

> **Most leech bites do not cause serious harm, though there will be extended bleeding from the wound once the leech is removed. Leech saliva contains an anticoagulant that prevents blood from clotting normally. Leeches can fall away on their own after feeding, but they can secrete some of their stomach contents into your skin before they go, which raises your risk of infection. "Leech saliva also has a natural anesthetic, so you don't feel them bite," says Cummins. "You may bleed for a little, but you won't die from this. Just keep any bitten areas clean and dry."**

Botfly larvae had burrowed under his skin—that was the lump above his eyebrow—and those were removed too. Doctors also treated his trench foot.

> Botflies are parasitic insects whose larvae grow under your skin. Infestation itself isn't typically harmful, but improper removal can burst the larvae, triggering an allergic reaction or secondary infection. However, as botflies grow under your skin, they can cause itchiness, pain, and a sensation of movement that can lead to insomnia. "This sensation would turn up the intensity for Yossi," says Giovanone. "It'll worsen everything, including his brain function. And not sleeping would contribute to delirium."

Ghinsberg went on to write a book, *Jungle*, about his ordeal. In 2017 the book was adapted into a movie of the same name starring Daniel Radcliffe.

HOW TO SURVIVE IN THE JUNGLE

If you're lost in the jungle, remember the acronym STOP: Stop, Think, Observe, and Plan. Orienting yourself can be challenging, but keep your brain engaged to avoid panicking.

STAY NEAR WATER, RIVERS, OR STREAMS. Ghinsberg kept changing the path of travel, first going upriver to find Kevin Gale, then downriver, then toward the mountains. That's unwise. If you're unable to determine the right direction in which to go, whatever direction you pick, keep moving on that path. You want to stay near water—streams or rivers—and keep moving downhill. Rivers will lead you out of the jungle, toward civilization. If you can't see a river, find higher ground and look for a depression in the terrain, which can indicate rivers.

FIND SHELTER. You need protection from the elements and predators, so if you can't find a cave or recess, you'll need to create your own. The easiest method is to lean a bunch of sticks or bamboo poles together and cover the outside with smaller branches, leaves, or other green foliage. If you can, build an elevated platform to get you off the ground and away from things like fire ants.

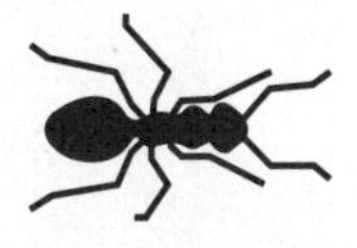

BEWARE FALLING TREES. A huge source of injuries in the jungle is when trees or tree branches fall. It can happen very quickly, especially during heavy rainstorms, as Ghinsberg learned. If possible, set up your shelter without any tree branches hanging over it.

FIND DRINKING WATER. Avoid drinking any stagnant or river water; there's too much bacteria in it. If you have no other options, you can boil this water or use purification tablets. You can also use leaves to collect rainwater and pour that into your water bottle. Make sure you're constantly drinking; you run the risk of dehydration in climates this warm.

LOOK FOR FOODS YOU RECOGNIZE. Plenty of fruits and plants in the jungle are poisonous, so don't eat anything you do not recognize. Depending on the jungle's location, there should be ample things that are familiar, including bananas, pineapples, oranges, mangoes, and avocados. Bamboo and palms are also safe to eat, and you can boil them to make them softer.

WATCH YOUR STEP. In rivers, rocks will be slippery and the current will be strong. If you have to cross water, grab two sticks for balance to help you remain upright. You can also use these to poke the water in front of you, to avoid falling into a deep hole you may not be able to see. On land, you can use walking sticks to test muddy puddles, to avoid getting caught in a bog, as Ghinsberg was.

STAY DRY. All wet skin is bad, but trench foot can be the worst predicament if you need to keep moving. "Put your feet inside plastic bags, then into your shoes," says Cummins. "That's better protection than your boots or shoes alone can likely provide, but once your foot starts to sweat, the bag will trap that moisture and your foot will become wet. Stop often to air out your skin."

BUILD FIRES. It's harder to build fires in wet conditions—but not impossible. After the rain stops, look for dead branches in trees; those are usually dry inside. You can also push over dead trees and find dry roots to use for kindling. You'll need more kindling than normal, so gather extra supplies. Use wet hardwood to form a platform over damp soil and place your kindling in the middle, then wet firewood around the platform's perimeter. Use waterproof matches or a lighter to ignite the kindling. As the fire grows, it'll dry out the wet logs.

MAKE A LIFEPACK. Bring as much gear as you can. Must-haves for jungle survival include insect repellent with DEET, flashlights, a knife, water bottles, plastic bags, waterproof ponchos, and a first-aid kit. The medical kit should contain antidiarrheal medication, sunburn salves, aspirin or other painkillers, antibiotics, waterproof bandages and tape, compression bandages, antiseptic wipes, compression bandages, and moleskin for blisters.

FORAGE NATURAL ANALGESICS. If you're lost and injured, and you've exhausted your medical supplies, you can use a variety of foraged foods (depending on your location) with anti-inflammatory and pain-relieving properties, per Palmisano. "Willow bark has a chemical called salicin, which is what they use to create aspirin," she says. "Chew it or boil it into a tea and drink that." Turmeric and tart cherry juice are both anti-inflammatory. "Peppermint oil is a painkiller," Palmisano says, "and hot peppers have capsaicin, which you can boil, then turn into a paste and apply as a topical analgesic." Last, rosemary has antibacterial, antiviral, and antifungal properties, and it can help improve blood circulation and immune system health. Before you venture into the wilderness, consult a field guide for which local plants might be useful and be sure you can identify them correctly. CAUTION: Do not consume or otherwise use wild plants unless you can ID them with absolute certainty!

STRANDED IN THE ARCTIC

5

LATE JUNE. EARLY MORNING.

Ada Blackjack stares at the corpse of Lorne Knight. The burly two-hundred-pound explorer died overnight, succumbing to a months-long battle with scurvy. The first thought the twenty-four-year-old Inupiaq woman has is horrifying: She's alone on a remote island in the middle of the Arctic Ocean.

"Scurvy can happen when there's a severe vitamin C deficiency in your diet. When our bodies don't get enough vitamin C, bleeding sores, tooth loss, and anemia can occur. The body also has a slower rate of healing from injuries, and if left untreated, scurvy can be fatal. "Vitamin C, or ascorbic acid, is needed for collagen production, which creates structure in our bodies, like blood vessels, tendons, and skin," says Matt Cummins, an emergency physician. "Without collagen, you literally can't hold yourself together. You'll die slowly as your body falls apart."

But it may not be the scurvy itself that kills you, adds Cummins. "If you're scurvied, your skin is breaking down and not healing, so you'll have open wounds. Those can get infected and that infection will be what kills you. You'll likely die of infection or internal bleeding."

Internal bleeding itself doesn't hurt because pain is not felt the same way with visceral organs. "Organs don't have pain receptors," says Beth Palmisano, a medical doctor who specializes in pain management. "We may fail to differentiate where the pain is coming from if an organ is bleeding. And you may not feel pain unless it's creating tension or pressure from a pocket of blood or fluid. For example, if your kidney is bleeding and the pressure from all that blood pushes on structures that have nerve fibers, that'll cause pain. But you can have a slow bleed in your intestine and you may not know immediately."

She shudders at the notion of facing a solitary existence in this godforsaken place, called Wrangel Island, with winter on the way and only a vague hope of rescue. She needs to focus if she's going to survive. First things first: *What do I do with Lorne's body?*

At five feet and barely one hundred pounds, Blackjack is too small to wrestle Knight's corpse to a grave. So she builds a barricade around his cot in the main tent, to keep away scavenging polar bears and foxes.

Polar bears can weigh more than seventeen hundred pounds and see humans as prey. Unlike brown or black bears, polar bears never hibernate; they hunt year-round. Nearly 30 percent of polar bear attacks on humans are fatal, bolstering the rhyme about how to deal with bears: If it's brown, lie down; if it's black, fight back; if it's white, say goodnight.

"We don't want those beasts sneaking up on us, right, Vic?" she says to the expedition's cat. He's the only living creature on this island around whom she feels safe. Vic stares at her blankly, then goes back to batting a half-dead mouse.

Over the next several days, Blackjack moves her living quarters to another of the camp's three tents—the supply tent—to avoid the smell of Knight's decomposing body. This tent needs some work, so she reinforces the sagging canvas walls with driftwood, constructs a sleeping platform, and makes a stove out of empty kerosene cans.

Each day, she prepares breakfast and dinner for herself and Vic: two meager portions of hardtack and a flour cracker, dipped in seal oil, to provide a few extra calories. It's not enough, but she's out of rations.

One night, as she and Vic gobble down their food, the cat looks at her expectantly. "I'm still hungry too," she says. Her eyes fall on Knight's rifle. "I guess I have to learn how to hunt."

* * *

Before setting foot on Wrangel Island's gravel beach in September of 1921, Ada Blackjack had never spent any time in the wild; she hated guns and had an extreme fear of polar bears. Spending two years on an Arctic island one hundred miles off the coast of Siberia was the last thing she expected to do.

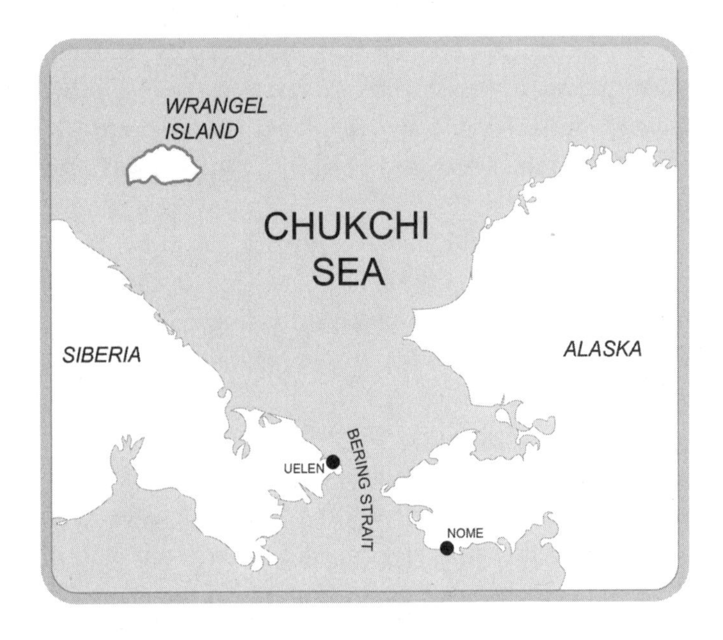

But the promise of more money than she'd ever seen led her to reluctantly accept a job as a seamstress and cook for an expedition of four white men who sought to claim Wrangel Island in the name of Canada. Though it gutted her, she'd left her five-year-old son, Bennett, stricken with tuberculosis, in the care of an orphanage back in her hometown of Nome, Alaska. The money would be enough to move to Seattle, where Bennett could get better medical care.

The expedition's first year on Wrangel Island passed uneventfully. But a ship sent to resupply rations couldn't make it through the frozen sea, and during their second winter, Blackjack and the explorers began running out of food. Desperate, three men took a team of sled dogs—so starved their ribs poked through their fur—and set off on a hundred-mile journey across the frozen sea, trying to reach Siberia and get help. Knight, too sick to move, was left behind with Blackjack. That was three months ago.

This morning, as she brings Knight's rifle outside, the sky is a beautiful shade of purple. In Arctic summers, the sun never completely sets, but it does sink near the horizon, giving the effect of dawn. The island looks vast and desolate, with not a tree in sight, only an unending array of jagged rocks and driftwood, with a mountain range in the distance. The fresh blooms of colorful wildflowers break up the bleak view, but Blackjack knows that with winter coming they'll be dead soon. And so will she if she doesn't figure out how to use the gun.

She puts the rifle and ammunition on the ground. She doesn't know the first thing about guns and loathes how noisy and scary the weapon is. She would cringe at the sound whenever the men used it. She fumbles to get it loaded, then

places the barrel on a large rock, for stability, and takes aim at a piece of driftwood in the distance. She slowly squeezes the trigger. It fires with an enormous bang, and the recoil knocks her down. Blackjack dusts herself off and tries again. And again.

For hours, she practices until she's used to the noise. *Okay, time to find something to eat.* She spots a bird perched on a rock, some forty feet away. She aims and, with her first shot, hits her mark. "Not bad!" she later says to Vic, as they eat her kill.

* * *

Three days later, Ada Blackjack's aim is getting better. As she eats a freshly killed bird, she knows this isn't sustainable. Birds offer too little in the way of meat, and besides, once winter comes, they'll disappear. She has no idea if the three other men ever reached Siberia, nor when—or even if—a relief ship is coming. She needs bigger game if she's going to survive another winter. That means hunting seals. Blackjack knows this is easier said than done; even the men, all experienced hunters, had trouble catching one. But she must try.

Seals like to lie on tiny ice floes that drift just offshore. Blackjack grabs the rifle and some ammo, then heads to the beach, a few hundred yards from camp. In the distance, a fat seal lounges on a floe close to the beach. Blackjack lies on her belly and crawls as close as she can without spooking the animal. Killing this seal would solve so many problems: Its meat would feed her, its blubber would fuel her lanterns, and its skin could be dried and turned into clothing to keep her warm in the frigid months ahead.

The seal is starting to nod off. Blackjack lies still, the rifle barrel propped on a piece of driftwood. She squints against the glare. Has the seal closed its eyes? She takes a breath and squeezes the trigger. A loud crack pierces the silent Arctic air. There's a spray of red as the seal's head jerks. Blackjack leaps up, lets out a whoop, and runs toward her kill.

But the dead seal is starting to slide into the ocean. *No, no, no!* she thinks, sprinting now. She splashes into the water and reaches the floe just as the seal's body falls into the ocean. She watches, helpless, as the current carries her prized meal away.

* * *

In the days since Knight died, Ada Blackjack has been getting good at building things. After losing the seal, she started building an umiak, a lightweight Inupiaq hunting boat that resembles a canoe. She needs this if she's going to keep hunting seals; getting the bodies from the water will be easier. Umiaks are essential to how Blackjack's people survive, but she never learned how to make one, so she's winging it. She hammers driftwood into the framing of the boat and sands it meticulously. It takes hours, but she's pleased with her progress.

"What do you think, Vic? Does this look like a boat?"

she asks. The cat, pouncing on nearby flowers, ignores her. "Well, *I* think I did a good job."

Traditional umiaks are covered with animal skins, which is better for waterproofing. But Blackjack doesn't have enough skins to make that work, so she grabs canvas from the supply tent and begins stitching pieces together to form a hull. It takes forever and her fingers throb, but with no nightfall during Arctic summers, she can work around the clock.

Two days later, Blackjack's umiak is finished. She beams at her creation, then wonders if it will actually float. Dragging it down to the water's edge, she sees a seal snoozing on an ice floe one hundred yards offshore. She carefully places the boat down, grabs her gun, and shoots the animal. Tug-

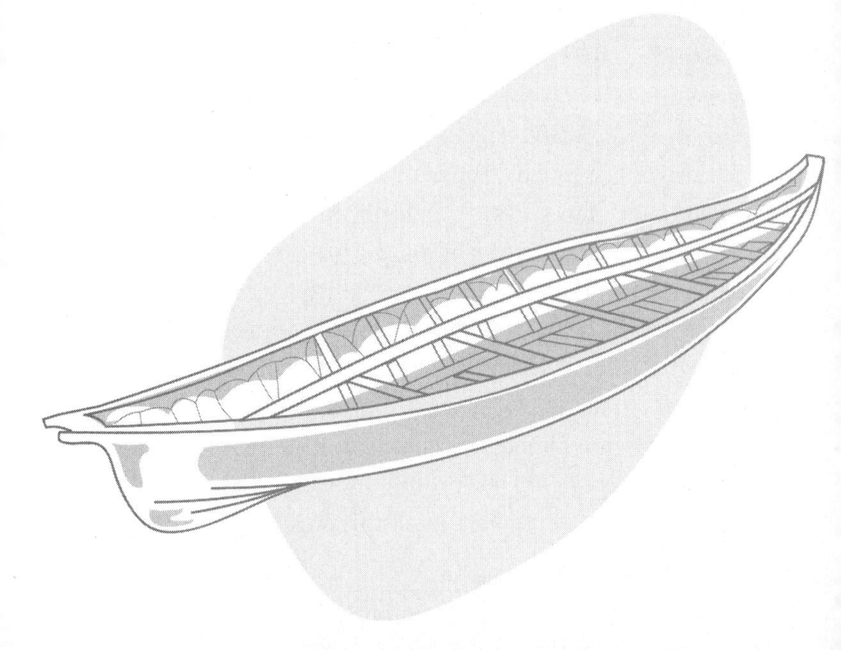

ging the boat into the frigid waters, she hops in. It stays upright. *Success!* Paddling furiously with two oars she carved from driftwood, she races toward the ice floe.

Just as the last one did, this dead seal is sliding toward the floe's edge. *Not again.* She rows faster, pleased at how well her boat slices through the small whitecaps. She finally reaches the floe and stretches an arm out toward the dead seal, grabbing its flipper. The seal outweighs her by fifty pounds but she finally lugs it on board. The umiak lists under the seal's weight, but it doesn't sink.

As she rows back to shore, for the first time since Knight died, she realizes: *I can survive here.*

* * *

Energized by a hearty dinner of seal meat, Blackjack goes hunting the next morning and bags yet another seal—this one on land. It's too heavy to drag back to camp, so she runs back and grabs an axe so she can butcher the carcass. But just as she nears the dead seal, she hears a distant growl. She turns to see a flash of yellow and white. It's a polar bear—still some two hundred yards away, but barreling toward her.

Blackjack quickly moves her rifle to aim at the beast, but her arms are shaking with fear. She can't afford to miss; she won't have time to reload, and that could be fatal. The polar bear closes the gap, a cub in tow. Blackjack can clearly see its curved black claws and powerful white teeth glistening in the sun. It lets out a roar as it draws closer.

Blackjack turns and sprints back to camp, diving inside her tent. She waits for the bear to arrive and maul her to death. Except nothing happens. She pokes her head out of her tent, scanning the horizon with binoculars. The bear and cub are on the shore, devouring the seal. *My seal.*

Enraged, she fires the rifle into the air, but the bears continue feasting. They're out of rifle range, and she doesn't dare move to within firing distance, so she watches helplessly.

She's fighting so hard to survive in this wretched environment. Yet it never feels like it's enough.

* * *

The weeks drag on, as July gives way to August. But Ada Blackjack hangs on with a sporadic diet of seal meat. She's been suffering headaches, stomachaches, hallucinations, and swollen eyes, but each day she keeps working to keep the camp running. Chopping wood. Baiting animal traps. Repairing the tents. She knows the ocean will freeze over soon, and ships won't be able to reach her. The thought of being stranded alone here for the winter terrifies her. Worse, she misses Bennett terribly and fears she'll never see him again.

People under extreme conditions of isolation may hallucinate. "The brain models the world around us and makes predictions so we don't waste energy trying to perceive everything freshly all the time," says Alex Sabo, a psychiatrist. "When we look at a landscape, our brain fills in gaps by imagining rocks or

trees in certain spots because those elements are around us." In the Arctic, sensory deprivation can begin with the harsh landscape, but the cold and lack of food can also cause the brain to activate nontypical parts. "This leads to experiencing hallucinations, which is the brain's predictive modeling gone wrong."

Freezing temperatures and the dryness of the Arctic air can also wreak havoc on eyes. "The Arctic has less moisture in the air," says Matt Cummins. "Your body is losing moisture faster as a result, even in your eyes. The cornea is not designed to be dry, so as it dries out, it's painful and itchy. The more you rub it, the more swelling, redness, and irritation occur." This is colloquially known as "winter eye" or "winter burn," and the only way to correct it is by replacing fluids. Drink a lot of water, and keep your eyes closed more often, to better distribute the eyes' natural lubricant, adds Cummins.

But dry air also increases the eye's sensitivity to light, even if it's overcast or cloudy. So snow blindness is also a concern. "The sun is super bright in snowy conditions, and the UV rays reflect off the snow more intensely," says Cummins. Your retinas can get sunburned rather easily—it'll be painful, and your eyes can swell. However, it's temporary and should heal in a few days without causing permanent damage. ”

One night, a tremendous windstorm hammers her tent, and Blackjack distracts herself by petting and cuddling Vic. In the morning, she steps outside her tent and surveys the damage: Supplies have been blown around, and the wind has tattered a few spots of the canvas tents, but it's all repairable. But the umiak is nowhere to be found. The wind must've blown it away. It's a devastating blow. Without the boat, she can't hunt seals.

She falls to the ground crying. Everything is exhausting. Constantly hunting for food. Constantly checking for polar bears. Constantly missing Bennett. Weak, she crawls into her tent, into her sleeping bag, and sobs. She remains in bed for three days. On the fourth day, Vic nudges her, meowing softly.

> **Loneliness can harm or even kill. When alone in an unrelenting environment, struggling to survive, humans have an increased risk of poor physical health. "Effective immune system function can decrease when people are isolated," says Sabo. "Organisms in a community feel safer, and protective systems are more active. If Ada Blackjack is isolated, in a chronic state of being vigilant, with no safety or security, her parasympathetic system—needed to stimulate growth hormones and help begin repairing damaged cells—can't engage as effectively."**

"I guess you're right," she says, rubbing Vic's chin. "Time to get back to work."

Over the next week, Blackjack builds another umiak. One day she's putting the final stitches on the canvas hull when there's a noise. It sounds like . . . a ship's whistle? She peers out into the foggy bay but sees nothing. *Must be a bird*, she thinks. The new umiak floats beautifully during a test run, and Blackjack looks forward to a hunting trip the next day.

But the next morning, before she can launch the boat, she hears that whistle again. The fog blanketing the bay obscures the sea, but Blackjack is sure it's not a bird. Suddenly, her swollen eyes make out a ship, emerging through the mist.

Blackjack leaps, runs, laughs, and cries. She's saved.

* * *

Ada Blackjack spent 703 days on Wrangel Island, fifty-seven of them alone. The three men who tried to reach Siberia with sled dogs were never found. After eight days aboard the rescue ship, Blackjack reached Nome, Alaska, and was reunited with Bennett and the rest of her family. Vic the cat was also rescued. Blackjack had symptoms of scurvy, though it cleared quickly upon reaching land.

> After her rescue, Blackjack credited Vic the cat for helping her stay alive and sane. Physiologically, pets do help us. "When you're in a bonded situation with a cat, your brain will release more oxytocin, which dampens fear signals," says Sabo. "It can help stabilize and decrease feelings of isolation. The oxytocin can also help your body move from a sympathetic state—when you need to be ready for fight or flight—and a parasympathetic state, so you can sleep and recover."

HOW TO SURVIVE IN THE ARCTIC

Ready to brave the Arctic wilds? Bring plenty of gear (that you've tested in warmer climates; cold temperatures make even the smallest tasks like tying a knot much harder) and ample food. Here's what you need to survive above the Arctic Circle.

STAY HYDRATED. Your body needs to be well hydrated to move blood and oxygen to your muscles and extremities. Make sure you're drinking water regularly to avoid dehydration.

EAT AMPLE CALORIES. Normally, adult women need 1,600 to 2,400 calories daily, while adult men need 2,200 to 3,000 calories. You need about 5,000 or more calories every day when in the Arctic to keep your body in homeostasis. And a chunk of those should be from foods high in fat. Burning all these calories is what'll help your body generate enough heat to stay warm, but the calories also properly fuel your body for the more demanding conditions. Struggling to hit that high minimum? Add a stick of butter per person, per day, to your meals.

GET PLENTY OF VITAMIN C. Most creatures make their own vitamin C, except bats, guinea pigs, and primates (including humans), says Cummins. To ward off scurvy and keep collagen production high, eat plenty of citrus fruits, which are high in vitamin C. "Diets of only meat, like Ada had, can lead to problems. Make sure you're eating things like bell peppers, tomatoes, and plenty of fruit," says Cummins. If you don't have access to fresh fruits and vegetables, use native foods like berries, kelp, licorice root, and mountain sorrel, staples of Inuit and Inupiaq diets.

WEAR FOUR LAYERS.
To stay warm, dress in loose layers. The air that's trapped between these fabrics will warm your body. Opt for wool layers, as they can help wick away sweat—you lose heat 240 times more quickly if you're wet.

USE A SILK BASE LAYER. "Polar thigh" is a painful condition that can form on your legs. It's effectively chilblains—itchy, swollen lumps or patches on your skin—and it doesn't take long for these patches to become blisters, then open wounds. Avoid polar thighs by wearing a layer of silk beneath your wool layers; it helps prevent abrasion. Should you suffer polar thigh, treat the affected areas with steroid cream.

PROTECT YOUR EYES. You want to shield your eyes from colder temperatures, but also the sun itself. Ski or snow goggles work best, and they can help keep your eyes warm and moist. But sunlight reflecting off snow or ice can intensify to the point where it can cause snow blindness. This condition can last for a few days and is extremely painful, and the damage to your eyes can be permanent. The solution is to tape your goggles so only a small slit remains to see through; it'll cut the blinding sun.

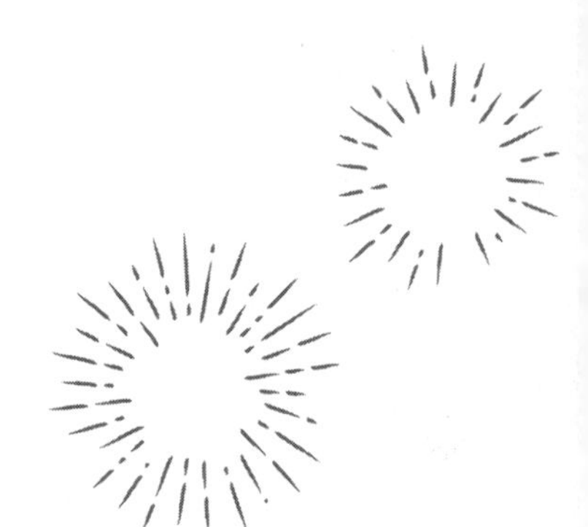

DETER POLAR BEARS. Polar bears have lousy eyesight and they rely on their senses of smell and hearing to hunt prey—including humans. Avoid Ada Blackjack's worst nightmare by making a lot of noise. Bear bangers—purpose-made bright, loud flares—can help keep polar bears at bay. Personal alarms that emit piercing sounds can scare polar bears too. Finally, dogs can detect bears sooner than humans, and their barking will often frighten them away.

KEEP EXTREMITIES WARM. Keeping toes and fingers warm throughout cold nights is essential for good blood flow. You can accomplish this by boiling enough snow to fill a one-liter insulated water bottle, then place that bottle at your feet in your sleeping bag. It'll still be warm in the morning, and you can put it inside your boots to heat those up before slipping your feet in.

BUILD AN IGLOO. Blackjack had a canvas tent, but the warmest long-term shelter in cold weather is an igloo. Build an igloo from blocks of snow, at least twelve inches thick, laid spirally for structural integrity and insulation. Scrape the interior walls smooth to help improve durability and ward off accumulating condensation. Igloos can last for months, provided the temperature outside is under thirty-nine degrees. Make a small ventilation hole at the top to avoid suffocation. Your igloo won't be warm, necessarily, but with proper clothing and a sleeping bag, you'll be able to live comfortably.

PACK MOUNDS OF SNOW UNTIL THEY HARDEN, OR CUT BLOCKS OF SNOW FROM THE DEPTH WHERE YOUR FEET STOP SINKING.

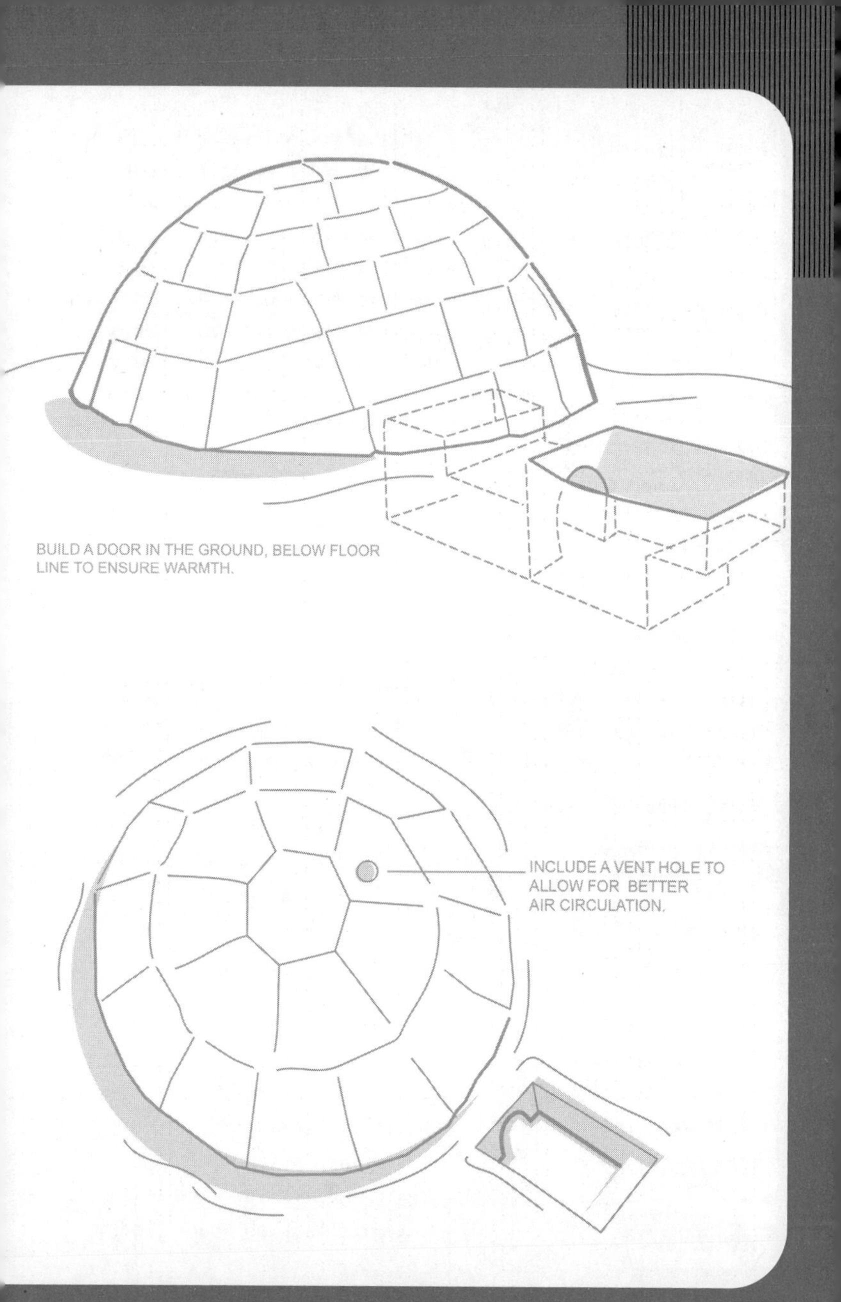
BUILD A DOOR IN THE GROUND, BELOW FLOOR LINE TO ENSURE WARMTH.
INCLUDE A VENT HOLE TO ALLOW FOR BETTER AIR CIRCULATION.

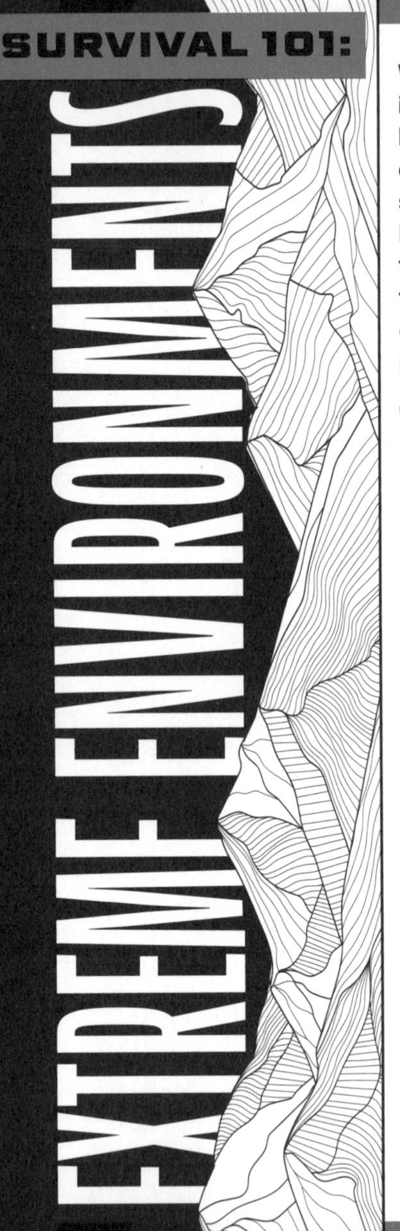

SURVIVAL 101: EXTREME ENVIRONMENTS

Whether you're floating in a life raft in the ocean, praying for rescue like Maralyn and Maurice Bailey, or caught in a mountain blizzard, shivering in a snow cave, like the Episcopal School group, a few basic tenets of survival apply. Commit these five basic tips to memory to greatly increase your chances of pulling through . . .

---

1. PREPARE PROPERLY

Regardless of the climate or terrain, all the survivors from these five stories had enough gear suited for the environment they were entering. By wearing situational-appropriate clothing, packing first-aid gear, and carrying basic survival tools, when the worst case becomes reality, you'll be able to power through relying on your supplies. Bring more gear than you think you'll need.

2. MAINTAIN YOUR TEMPERATURE

The human body functions properly within a relatively narrow temperature window: forty-one to ninety-five degrees Fahrenheit, though we can tolerate hotter climates if the humidity is below 50 percent. When we get too hot or too cold, we rapidly lose cognitive function, a dire necessity to continue the fight for life. Our organs and circulatory systems also begin to fail when hyper- or hypothermia begins, starting a countdown clock to potential death. Make sure you're always able to get warm or cool.

3. HYDRATION IS EVERYTHING

You can survive days or even weeks without food—but you won't make it much past three or four days without water. Ed Rosenthal didn't have much hope left in the California desert after his water supply was exhausted. Always bring ample water reserves, at least one liter for every hour you plan to hike, and more if you're hiking in hotter temperatures. Also do plenty of research on how to find or produce water from your surroundings before setting out.

4. PROTECT YOUR SKIN

Our skin is our greatest barrier against infection and disease. It's also more delicate than we realize. Cuts, insect bites, sunburns—a variety of issues—can penetrate our skin, allowing bacteria to enter. And once any infection starts, it becomes a much more difficult struggle to keep going. Make sure your first-aid kit includes lots of skin protection—antihistamines for hives and itching, antibacterial cream for cuts, sunscreen with a high SPF, and hydrocortisone cream for rashes are all essentials.

5. STAY PUT

It's easier to find lost, injured, or stranded victims if they're stationary. If you're on the move, any search parties will have to cover larger swaths of ground to locate you. Whenever possible, stay in one spot. (This can be harder to do in an ocean current, but drift socks or drift anchors can be used to reduce the rate of a vessel's drift.)

PART II

ATTACKED BY ANIMALS

Facing off with a wild animal is terrifying. When you're up against a charging brown bear, a hungry crocodile, or a venomous snake, these heart-pounding situations trigger our fight-or-flight systems and our bodies instinctively react in a surprising number of ways. Most animal attacks are brief, but getting medical help after you've been gravely wounded can be as challenging as surviving the attack itself . . .

BITTEN BY A SHARK

APRIL 29. LATE AFTERNOON.

Dusty Phillips beams as his girlfriend, Leeanne Ericson, arrives at the San Onofre beach near San Clemente, California. Ericson's shift as a bank teller has just ended, but there's enough daylight left to get in a little ocean time on this Saturday afternoon. Phillips secures the rubber leash of his yellow surfboard around his right ankle and looks at Ericson, who's tugging on her wetsuit.

"Faster, slowpoke," Phillips teases. "I'm missing all the good waves." Ericson smiles and grabs her swimming fins. She's thirty-five and, though she's not a surfer, she takes every chance to go for a swim in the ocean.

"Ready," she replies. The two make their way into the water and swim out to the breakers, where Phillips waits for a solid wave to catch. It's a beautiful day, he thinks. As the sun drops toward the horizon, its light bounces off a pod of dolphins as they jump above the waves. The wind kicks up, and the water's a little murky, but the waves are decent, and Phillips whoops as he catches a solid set.

He's taking a breather less than one hundred yards from shore when Ericson surfaces nearby, treading water. She flashes him a wide grin, and he pats the top of his board, an invitation to come join him. She swims over and hoists herself onto the board, and the two sit in the surf, facing each other, legs dangling below.

Suddenly, there's a dark streak of motion to Phillips's left. A sea lion, breaching. To Phillips, it looks panicked and confused. Ericson clocks this too. "You know what that means, right?" She gives him a smirk. "Probably running from a shark on the hunt."

The couple exchange a few shark jokes and chat about how their days went, when Ericson jerks her leg and locks eyes with Phillips. "Did you just kick my fin?" she asks. "That's not funny!"

But Phillips doesn't respond; his eyes are fixed on a huge wave rolling in from behind Ericson. "Gotta go!" he says, as he moves the board to displace her and catch the wave. As Ericson plops into the water with a giant splash, Phillips paddles away.

But then a piercing scream fills his ears, coming from behind. Just as quickly, it stops, as if it's gone underwater. *Cute*, Phillips thinks. *Is she pranking me?* He glances back, expecting to see Ericson laughing.

Except she's gone. The ocean's surface where she was just moments ago is completely glassy. Not a single ripple.

Phillips has to give it to her; ducking under after screaming is a good move. He waits for her to pop up. But she doesn't surface. A few seconds pass. The stillness sends a shiver down Phillips's spine. *Where did she go?* He dives underwater, stretching the coiled leash that connects him to the board. In the murky water, he strains to see anything. Nothing.

He surfaces, takes in air, and looks around. Ericson's still nowhere to be seen. He dives again, deeper this time. It's too hazy to see, and his lungs burn from a lack of oxygen, driving him back to the surface. Shaking water out of his eyes, his head swivels around. No Ericson. But then, out of the corner of his eye, a blur, maybe twenty-five yards away.

Is that . . . a tail? Phillips stares in disbelief. It is, and it's thrashing, sending frothy water in all directions. It's a shark. And not just any shark. Judging by its size—at least ten feet—it's a great white.

And it's feeding on his girlfriend.

Shark attacks are very rare—the odds of it happening are one in 3.75 million—and only a very few are fatal. There were forty-one recorded shark attacks in the United States in 2022, and only one of those was fatal.

Unfortunately for Leeanne Ericson, she is now in that rare company of attack victims. A shark has sunk its rows of razor-sharp teeth into the back of her right thigh.

The pain is otherworldly, the pressure unbearable. Impossibly, it gets worse as the shark's jaws clamp down and she feels herself being dragged beneath the surface. Her first thought as the shark descends is: *Well, it finally happened.*

She knows there's been an increasing number of shark sightings off this beach. She also knows that this isn't a test bite—something sharks do to see if what they're chomping on is indeed prey. It's an attack bite, to kill her.

As the shark pulls her deeper, the faces of her three young children flash in Ericson's mind. *Fight to stay alive for them*, she thinks. Her leg still caught in the shark's mouth, she reaches down and moves her fingers over the shark's hard, rough skin. It's like armor. There must be a soft spot somewhere—a place she can attack.

In seconds, the shark has pulled her to such a depth that the sunlight can't reach. Not only is it dark, it's quiet; there's no noise underwater. The only thing she hears is her brain, screaming that she needs to take a breath, immediately. Her fingers are still searching the shark's tough skin. Finally, she feels something soft, like Jell-O. *Must be the eye.* Instinctively, she jams her fingers into the area, as hard as she can.

"If a shark drags you underwater and leaves you unable to surface for air, your body's reflex is a laryngeal spasm, which closes off your airway, meaning you can no longer breathe. "Leeanne was struggling underwater, so she's exerting more energy and her oxygen requirement goes up," says Deepak Sachdeva, an emergency physician. "Whatever oxygen she had in her lungs before she got pulled down would be quickly used up. When there's no fresh oxygen, you're quicker to develop symptoms of tissue damage," he says. Since she ingested water into her lungs, that will also cause damage, says Sachdeva. "The sacs inside your lungs are called alveoli, and when those are filled with water there will be cellular damage, so your lungs won't function well," he explains."

The pressure on her leg instantly vanishes, and now she's floating upward. It's growing light again as she nears the surface. Her lungs are burning. *Air. Now.* Then her head pops up. She gulps in air and cries out for Phillips, who's twenty feet away. He paddles madly toward her and then hoists her onto the board.

"I don't want to lose my leg," she pants.

"You won't, babe," he replies, wrapping his arm around her hip. He squeezes hard, trying to stop the bleeding. It barely slows.

Ericson looks around, worried the shark is going to return to finish the job. Blood pours from the wound, staining the yellow surfboard and turning the water a dark crimson.

> **What happens to our pain thresholds when we're bleeding out? "You're going to have a huge endorphin release," says Beth Palmisano, a medical doctor who specializes in pain management. "Natural opioids will surge, which will help relieve the pain. You're more able to deal with other things in the moment, like being able to breathe, as Leeanne did, instead of focusing on your blood loss and the pain from the wound."**

Won't this attract more sharks? But she doesn't see any fins, so she looks down to see how bad the wound is. A huge chunk of muscle and flesh is missing, from her butt to her knee, and bone is visible. To her, it looks almost surreal, like a cartoon image of a bite wound.

> **When Ericson looked at her bloodied leg and thought she was looking at a cartoon, what was really happening was dissociation, says Anthony Giovanone, a doctor of osteopathy and a psychiatrist. "We dissociate for survival; we have to continue in moments of trauma," he says. "Trauma memories can be choppy because of the way the hippocampus is affected, the portion of our brain that plays a large role in learning and memory. Instead of coding a bad visual, the mind thinks it should fill it with a more palatable image than what she actually saw, in this case being a cartoon."**

Exhausted, she puts her head down on the board and tries to focus on breathing. Phillips begins paddling her to shore, screaming for assistance. The wind carries away his shouts; no one on the beach reacts. Kids keep playing in the shallows; surfers keep catching waves. Ericson is breathing, but it doesn't feel like enough oxygen is getting into her body.

She looks at Phillips and says, "I can't breathe," over and over. It feels like if she stops saying it she'll pass out and die. So she whispers the three words repeatedly as Phillips pushes her to shore. Each time she says it, she can feel him kicking harder. When they get within fifty yards of shore, Phillips's screams are finally audible, and two beach bystanders rush into the water to help get Ericson to the sand.

* * *

The bite the shark took out of Leeanne Ericson's leg was enormous. It removed more than one foot of flesh and muscle from her leg. The damage was so severe that inches of bones were visible. Ericson lost so much blood that by the time she reached shore, she was pale gray and Dusty Phillips was sure she was going to die.

"Because of its rows of razor-sharp teeth, when a shark bites and shakes, it can cause massive tissue damage and loss. In Leeanne Ericson's case, the bite was so deep it exposed bone and severed the sciatic nerve, which runs through the back of the thigh and connects the nerves of the leg to the spinal cord. "Sciatic nerve damage is extremely significant," says Sachdeva. "Even with reattachment, nerves don't regenerate well, and you need your sciatic nerve to control the lower parts of your leg, your movement, and sensation." Skin grafts are possible, though muscle regeneration is nominal, and muscle reconstruction is difficult and has high risks. "You also want vascular experts to assess whether the blood vessels are salvageable," says Sachdeva. "If the bite had severed her leg, specialists often wouldn't attempt to reattach it. There's just too much damage.""

Waiting for the ambulance, Phillips tried to tie a tourniquet with his rubber surfboard leash, but it wasn't effective; there wasn't enough flesh left on her hip to stanch the flow. Paramedics arrived, and within fifteen minutes of the attack, Ericson was in an ambulance on the way to a medical helicopter.

> Ericson's pale gray skin was a sign of massive blood loss. When the body is bleeding heavily, it redirects all blood to the chest and head, to keep the brain and organs going, leaving you with gray or white skin. "She was very close to death," says Sachdeva. "When you can't control the bleeding with a tourniquet or direct pressure—and her boyfriend tried both—you can't do anything else on the scene except get to trained medical professionals as fast as possible."

The helicopter medics induced a coma, and doctors kept her unconscious for a week while they operated on her. Her inability to breathe was caused by the water she'd inhaled when the shark was chewing on her.

Ericson spent nine weeks in the intensive care unit, had eight pints of blood transfused—enough blood to replace the average person's total blood volume—and underwent eight separate surgeries. Doctors debated amputating her leg since the sciatic nerve, which connects the leg and central nervous system, was severed.

> When replacing your entire blood volume, there's a concern about your blood's ability to clot, says Sachdeva. "If you're just getting blood but not the components that help blood clot, such as plasma and platelets, you risk increased bleeding,"

he explains. "In massive transfusions, such as this, they now transfuse plasma and platelets to help the body."

But a specialist reattached the nerve, and surgeons grafted skin from her left thigh to replace some tissue on the injured right thigh. So many antibiotics were required to stave off infection that the cost of medicine alone was more than thirty-six thousand dollars.

With such severe injuries, the risk of infection during the healing process is extremely high, hence Ericson's need for massive amounts of antibiotics. "The most common infection here would be polymicrobial," says Sachdeva. "There are multiple bacteria at play—from the shark's mouth during the bite, what's in the seawater, and what happens on land after the attack." Skin grafts increase your risk of infection—as does spending prolonged time in the hospital. "Hospitals have their own special germs, and you risk what's called hospital-acquired infections. Nine weeks in the ICU could definitely do it. In a weakened and down state, having lost a lot of blood, fighting infection would be harder for Leeanne," says Sachdeva. "If the infection takes hold, it could progress to sepsis, which can kill."

Remarkably, a few weeks after the final surgery, Ericson was able to lift the damaged leg, shocking her doctors, who believed she would never regain motion. Months later, she began walking, though she still requires a special brace. She has limited feeling below the knee and cannot feel anything below her ankle. The lifelong ocean lover can never look at the ocean the same way again and required medication for PTSD. She still gets very worried when her children swim in the sea.

> **Post-traumatic stress disorder is when your brain and body stay in survival mode long after the event. "You're stuck in the past, and constantly secreting cortisol, the stress hormone," says Giovanone. "With constant cortisol in our system, it leads to a number of bad things down the road, including heart disease, coronary artery disease, and depression."**

HOW TO SURVIVE A SHARK ATTACK

In Southern California, great white sharks are swimming near surfers 97 percent of the time, according to a recent study—and surfers are rarely aware of them. Here's what you need to know about sharks and when they attack.

LEARN WHERE SHARKS LURK. Shark attacks are most common near the shore. When low tide arrives, sharks can be trapped between sandbars, which tend to be around the same areas that surfers are trying to catch waves. Be vigilant about looking around if you're in a sandbar area.

STAY IN A GROUP. Sharks are more likely to attack people who are alone, so stick close to fellow swimmers or surfers. If you're bobbing in the ocean, try to situate yourselves back-to-back, so everyone is facing outward, to get better eyes on approaching sharks.

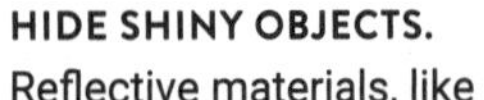

HIDE SHINY OBJECTS. Reflective materials, like watches or jewelry, are a glittery invitation for a shark beneath the surface. Sharks are attracted to colors like yellow and orange because they resemble the light that bounces off fish scales.

DON'T THRASH. Sharks have what's called lateral line systems, a sequence of fluid-filled pores that stretch the length of the animal's skin. These receptors are highly sensitive to detecting motion and pressure changes in the water, even from hundreds of yards away. If you see a shark, the less you move, the better. Leeanne Ericson's giant splash off Dusty Phillips's board likely got the shark's attention because it may have thought she was a sea lion it had been chasing.

KEEP EYE CONTACT. If you're in the water and you see a shark, look directly at it. Maintaining eye contact lets the shark know you're not easy prey that it can sneak up on. Don't look away from the shark as you slowly swim to shore.

RECOGNIZE COMMON SHARK ATTACK TYPES. There are three common kinds of attacks. The first is a drive-by attack, where the shark takes a single bite, realizes that you're human and not normal prey, and doesn't return; a bump-and-bite attack, where the shark brushes against you before returning to fully attack; and a sneak attack, where there's no advance warning—just a barrage of biting. The latter is what Ericson suffered.

IF BIT, FIGHT VIOLENTLY. Take a page out of Ericson's book and aggressively attack back. While the most sensitive parts on a shark are its eyes, nose, and gills, ferocious blows anywhere on its body may be enough to startle the shark into leaving you alone. If possible, stick your hands into its gills and pull hard. The tissue inside the gills is sensitive, and it will hurt the shark, causing it to release.

STOP THE BLEEDING QUICKLY. If you have serious damage after being bitten, act fast to stem flowing blood. If the bitten area is a limb, apply a tourniquet very high up on the limb, close to your groin or armpit. "Applying a tourniquet properly is painful," says Sachdeva. "It must be tight enough to stop the bleeding." If the bleeding only slows but doesn't stop, you can try applying another tourniquet below the first," says Sachdeva. Once it's on, you have limited time to get help before you risk losing the limb. "We call it the golden hour," says Sachdeva. "You need this injury treated in the first hour."

MAULED BY A BEAR

7

FEBRUARY 6. AFTERNOON.

Bart Pieciul breathes in a lungful of the ten-degree winter air. It stings, but he loves it. He's hiking up a steep slope through calf-deep snow outside of Haines, Alaska, along with two friends. The morning sun streams through hemlock and spruce trees, hitting his face, warming him slightly. Pristine days like this are why the thirty-eight-year-old Pieciul moved here from Poland.

As a teenager, he'd seen snowboarding videos out of Haines and instantly fell in love with the tiny Alaskan community, population twenty-seven hundred. The snow-capped mountains, the deep-blue water, the isolation—it all called to him. Even after eight years here, part of him still can't believe he's lucky enough to call this vast outdoor playground home.

He stops to adjust his shoulder straps. He's carrying ski poles and a splitboard—a snowboard that separates into skis—on his back. A hundred feet ahead of him up a ridge is his buddy Jeff Moskowitz, a thirty-three-year-old forecaster with the Haines Avalanche Education Center. Besides his own splitboard, Moskowitz is carrying a backpack stuffed with survival gear—what their friend Graham Kraft called "a clown car of equipment" this morning in the parking lot. Pieciul chuckles now at the description. He looks past Moskowitz but can't see Kraft, who's still farther ahead, out of sight up the ridge. Once they're at the top, the three men will ski back down while Kraft's dog, Tsirku, runs after them.

It's been slower going than Pieciul would like. The mountainside is blanketed with several inches of fresh powder, but the hard icy base beneath makes for a slippery ascent. Still, they're making decent progress. Pieciul is concentrating on picking the best line through the snow when a flash of brown catches his eye.

He squints. About fifty feet ahead of him, a patch of brown fur stands out against the dazzling snowscape. It's moving. At first Pieciul thinks it might be a porcupine.

But the patch of fur is quickly growing larger. What was the size of a soccer ball a second ago is now bigger than a boulder. In a flash, Pieciul realizes what it is. *Shit, shit, shit. That's a brown bear.* His friends must've passed right over its den and woken it up. Adrenaline surges through Pieciul's body. He stops in his tracks as his mind races.

The brown bear snorts as it clears itself from the snow. Pieciul knows brown bears can weigh up to sixteen hundred pounds; while this one isn't quite that large, it's big enough to make his mouth go dry with fear.

The area around Haines is home to some three hundred brown bears, roughly one bear for every nine citizens. The year before, there'd been a record number of run-ins between the human and bear populations.

But most encounters are in town, where bears sometimes come in search of food. In town, they're easily spooked and almost never attack humans. This is different: Pieciul is in bear territory. And this animal could kill him. He remembers what Haines residents are told time and time again by rangers: *Don't run. Bears chase things that run, and they can hit 40 mph.* What Pieciul decides now could determine whether this is simply a bear encounter or a bear attack.

He strains to see beyond the bear, now wagging its head, shaking off the doldrums of hibernation. Moskowitz is too far ahead, and he can't see Kraft at all. *No one can help you. Look big and let it know you're not prey.* Pieciul stands as big as he can and stretches his arms wide, over his head, moving them deliberately but not quickly.

"Bear, go away. Go away, bear," he says in a firm speaking voice. Yelling can startle a bear, provoking an attack. But talking sternly and clearly lets the bear know it's dealing with a human, which is usually enough to make it back off.

But this bear is standing its ground. The bear snarls, a fog of white breath erupting from its mouth in the cold air. Pieciul can make out its long yellow teeth, which send another jolt of terror through his body. He realizes his voice may not be enough. He starts to repeat his words when there's another flash of brown from behind the bear.

It's a cub, yawning and sniffing the air. This is a worst-case scenario. Mother bears will do anything to protect their young.

The bear begins plodding toward him, her claws gripping the icy snow with ease. She's advancing at a glacial pace but advancing nonetheless. Forty feet. Thirty feet. Pieciul gulps,

inching backward toward a grove of trees, and raises his voice, a last-ditch attempt to avoid an attack.

"Go away, bear! Go away! Leave me alone!"

The bear pauses for a moment, almost contemplative. She shakes her head one last time, as if saying "no." Then she charges.

* * *

Jeff Moskowitz isn't sure he heard it right. *Did Bart just say "bear"*? Moskowitz turns, and his eyes widen as he takes in the tense scene unfolding on the ridge below him. There, in the clearing, his friend is cornered, trying to talk down a sizable brown bear who's guarding her cub. Moskowitz is one hundred feet above Pieciul and fifty feet above the bear; he won't be able to help. As the bear snorts loudly, it's clear to Moskowitz the beast is angry.

A split second later, he sees the bear charge Pieciul.

The bear reaches Pieciul, a snarling mass of brown fur and rippling muscles. Moskowitz sees his friend try to shield his face, and the bear chomps down on Pieciul's left wrist.

Bears go after the face and upper body first, as Pieciul experienced. Expect a paw to the face, or a bite to your neck. With claws between two and four inches long, brown bears (including the dreaded grizzly, a subspecies of brown bear) can do a whopping amount of damage in a single swipe. Your eye socket can be broken, your skin or eyelids can be torn off,

and your nose or ears can be severed. "We think of trauma as either blunt or penetrating," says Deepak Sachdeva, an emergency physician. "Blunt is a whack to the head with a bear paw, while penetrating is the bear biting you and tearing your tissue. You can die from both, but penetrating injuries are more worrisome because of the potential for rapid blood loss." Sachdeva says facial injuries, such as a broken eye socket or having a nose ripped off, may be horrific and can cause permanent disfigurement, but they won't immediately kill you.

Pieciul is six-foot-two and two hundred pounds, but the bear lifts him off his feet as if he's a toy. Moskowitz hopes the bear will toss Pieciul aside, then retreat to her den. But the bear doesn't relent. Instead, she swings Pieciul violently, never releasing his wrist. *She's going to kill him*, Moskowitz realizes, but he's too stunned to move.

As the mother bear continues to shake Pieciul, the two start sliding down the mountain, picking up speed on the incline. Still the bear won't let go of Pieciul, even as they approach a drop-off.

And then they vanish out of sight.

Time slows. Pieciul feels like he's outside his own body watching this horror unfold.

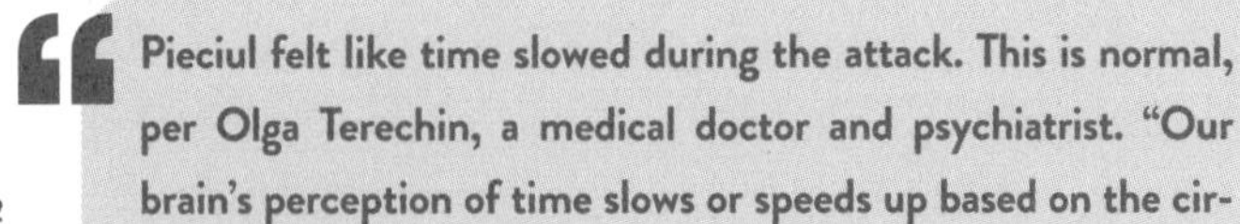

Pieciul felt like time slowed during the attack. This is normal, per Olga Terechin, a medical doctor and psychiatrist. "Our brain's perception of time slows or speeds up based on the cir-

cumstances. In extreme stress, one leading theory for why time may feel slower is that our information processing speeds up; we're thinking faster, which creates a perception that time is moving slower," Terechin explains. "Another theory is that your brain is trying to register as many of these horrific details to avoid this type of situation in the future. Because your brain is recording so many rich moments, when you recall these events, it feels that time was passing slower," she adds.

As the bear shakes Pieciul, the fingers on his left hand flop limply. Seeing this, the gravity of the situation washes over him. *Is she going to let me go? Is she going to hit an artery in my wrist*? Then he and the bear are sliding, flying over the ice. His helmet bounces along the ground. His backpack and splitboard are still attached. He closes his eyes. *Maybe the splitboard will keep her away from my neck.*

Always protect your head and neck, warns Sachdeva. "If a bear hits your head hard enough, you can have instant skull fractures and brain bleeds," he says. "Those are things that can cause immediate death."

The slide ceases but something's still very wrong. He tries to breathe but can't. He opens his eyes and immediately wishes he hadn't. The bear's snout is millimeters from his face. He feels spittle dripping onto his face. The animal is on top of him, crushing his chest and torso. He's pinned to the ground, unable to free a limb to protect himself.

Trapped, he watches in terror as her mouth opens wide. Her hot breath envelops him, her teeth glistening. Flaring her nostrils, she leans in and bites the left side of his neck, tearing away his flesh and part of his ear.

> **Bears often bite victims by the neck and then shake them, like a dog with a chew toy. This instinctive motion is an attempt to snap the victim's neck. "Neck injuries are bad because your spinal cord, major blood vessels, and arteries are all there," says Sachdeva. "Nerves from the brain go down the neck to control your breathing, and one strong bite can damage those, leaving you unable to breathe." As the bear shakes you violently, that whiplash movement can cause damage too. "If your spinal cord breaks, you risk paralysis. Or the shaking could damage your airway and affect your breathing, and potentially kill you."**

Play dead. No matter how bad this gets, play dead. Pieciul closes his eyes and wills his body to go limp. *If she thinks you're dead, you're no longer a threat.* After biting his neck, the bear moves down his body. Over and over, her teeth sink into Pieciul. His right hip, his buttocks, his right arm. The pain is unreal, but he cannot scream or move. He knows it will only make things worse.

This is what death feels like. Please just let it be quick.

> Bart Pieciul knew to play dead as the bear mauled him. This is different from fight or flight or freeze, since willing yourself not to move while being attacked is a conscious decision. What's happening physiologically in that exact instance? "This is an outlier of a scenario for me to explain," says Terechin. "Initially, Bart's amygdala is screaming to fight or flight, but his prefrontal cortex knows that's not right and is trying to respond more rationally, telling him to be still." However, as the bear starts tearing into his neck, "his stillness could have converted to freeze. Shocking pain from the bear digging into his neck could trigger a freeze response," says Terechin, leaving him unable to do anything other than lie there.

* * *

As Jeff Moskowitz straps on his skis to race down to Pieciul, he sees a blur streak by. It's Graham Kraft, who must have also heard Pieciul's shouts. Tsirku trails her owner, barking wildly. Moskowitz takes off behind them. Broken tree branches and discarded ski poles lead to the crest where he'd last seen his friend and the bear. He nears the edge and peers over.

Pieciul lies in a crumpled heap, the white snow around him turning crimson from his blood. There's no sign of the bear.

He and Kraft rush to their friend's side and pop out of their skis. Pieciul's breathing, but it's a dire situation. His left hand is at a grotesque angle from his wrist, broken severely during the bear's shaking. Gaping puncture wounds from the bear's long canines dot his neck. Blood drips from the tip of his nose, staining his beard and pooling in his jacket. Moskowitz is still surveying the full extent of the damage when Pieciul starts mumbling.

> Brown bears, like the one Pieciul encountered, have an average bite force of 1,160 pounds per square inch. Some human bones can break under as little as one hundred pounds per square inch, so if the bear is biting, odds are high you'll have a broken bone. "It depends on which bones we're talking about, but your best bet is to immobilize the area, like an arm or leg, as Bart's friend did," says Sachdeva. "The more things move, the more it can cause damage. When you remove any jostling, it also helps with the pain."

Moskowitz and Kraft lean close to their friend's stained face.

"What's that, buddy?" Moskowitz asks.

Pieciul's voice is slurred, but his friends hear the word loud and clear: "Helicopter."

Moskowitz's eyes meet Kraft's. Both understand the request; Pieciul's injuries are too extensive for Haines's small medical clinic. If he's going to survive, Pieciul needs a medical helicopter to evacuate him to Juneau's trauma hospital ninety miles away. And he needs it now.

"We're on it. You're going to be okay," says Kraft. "Mosky, sat phone. Call for help."

Moskowitz flings his backpack on the ground, unzips it,

and plucks out what looks like a clunky old cell phone. It's an inReach, a satellite device with an SOS button that sends GPS coordinates to emergency crews. His fingers shaking, Moskowitz presses the button, then texts his girlfriend as a backup, in case the SOS signal doesn't get through.

Pieciul moans. Moskowitz looks up from the phone at Kraft, who's examining Pieciul's wounds. The most dire is the wrist; through the mess of torn skin and muscle, Kraft can see broken bones. He's bleeding from other wounds hidden under his layers too, but Kraft doesn't want to strip his clothes off to check them. Given the frigid temperatures, hypothermia stands as much of a chance at killing Pieciul as his injuries do.

> **If the attack occurs in winter climates, beware of hypothermia. If you're somewhere very remote and help is hours away, as it was for Pieciul, keeping your body warm is vital. Bring extra layers and a packable sleeping bag to use while you wait for help.**

"The passwords to my computer are written on a Post-it that's taped under my desk," Pieciul whispers.

"Stop, Bart. You're not going to die," Moskowitz says. But he's not sure whom he's trying to convince more—his friend or himself.

Pieciul ignores him. "Tell my mom I love her. My social security number is . . ."

Kraft cuts him off. "Bart, we need to get your wrist stabilized. It's going to hurt like hell, but we gotta do it, buddy. Jeff, I need your backpack."

> "If the broken bones are compound fractures—where bone is penetrating tissue—that's a grave situation. "The bone itself may not be visible; you may look down and see a small cut or puncture," says Sachdeva, but underneath the bone is severely broken. "If the bone is sticking out of your skin, that's serious tissue damage. We call these open fractures, and they're problematic due to outside bacteria coming in and causing infections." Sachdeva advises cleaning these types of wounds with water and immobilizing them with a splint."

Moskowitz nods and unloads his pack. Out comes several heavy down layers, a sleeping pad, space blanket, a thermos of warm water, a first-aid kit, and a bivouac sack, a light cover that fits over a sleeping bag like a one-person tent. Kraft starts rolling up Moskowitz's now-empty, frameless backpack tightly, to transform it into a splint.

Moskowitz covers Pieciul with some down layers, the space blanket, and the bivy sack. Pieciul's face is draining of blood and his teeth are chattering more. Unsure how long Pieciul has, Moskowitz worries that the distress call didn't work.

He checks the inReach. His girlfriend has responded; she's relayed their GPS location to an emergency crew.

"Help is on the way, Bart. Hang in there," Moskowitz says.

Kraft kneels beside Pieciul, makeshift splint in hand. "All right, here we go," he says, and starts to shimmy the splint under his friend's mangled arm. Pieciul's screams are so agonizing that Moskowitz balls his hands and digs his fingernails into his palms as a distraction. Finally, after a few minutes, Kraft has tied off the splint. Pieciul pants, glassy-eyed, and sits up, accepting some water from Moskowitz's thermos.

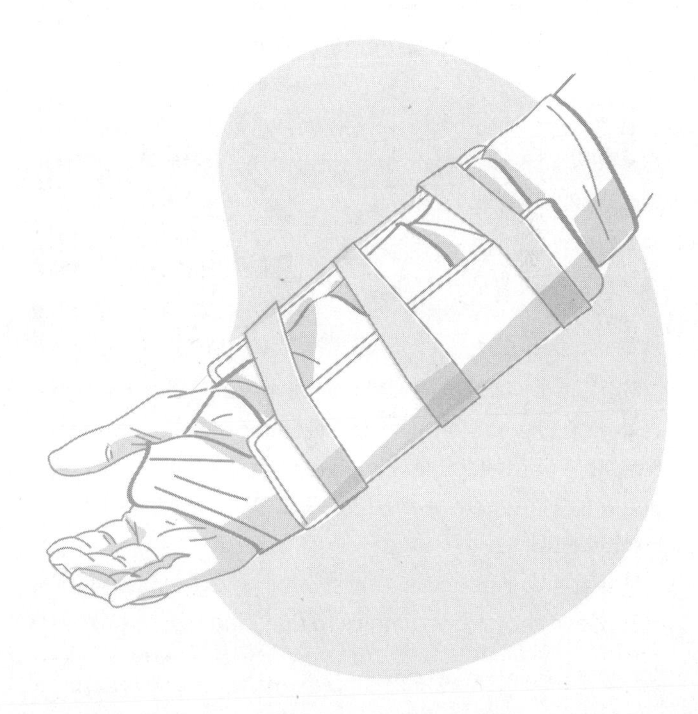

> Don't try to set a broken bone in the field unless the fracture is affecting blood supply. "When something breaks, blood vessels get disrupted or injured, hindering the blood supply to that part of the body after that point of injury," says Sachdeva. "If there's no pulse to the area below the injury—for example, in your foot if your femur is broken—you need to restore the blood flow quickly." Other indicators that blood flow is compromised include the skin looking blue, gray, or white, or feeling cold to the touch. The longer the area is without blood, the greater the risk of tissue death.
>
> But if you're faced with an extremity not getting blood circulation and help is hours away, you should try setting the fracture in the field. "It'll be difficult for a nonmedical person to pull a bone out and set it," says Sachdeva. "The aim is to pull the bone out longitudinally, as far as you can, then allow it to fall back into place naturally." The force required to set a bone can vary, but as you'd expect, "pulling bones back into place is very painful," says Sachdeva. "If you don't have to, don't do it."

Moskowitz tries to keep Pieciul talking. Slurred words are a sign of hypothermia, and his friend's sentences are increasingly unintelligible. He glances at his watch. It's only been a half hour since the incident. They need to get him warmer. Now.

Moskowitz rips apart a notebook and fumbles with a book of matches. He scrambles to find kindling. As he works to get the fire started, Kraft gingerly crawls next to Pieciul. "Ready to be the little spoon?" Kraft says, an attempt at levity that draws a laugh from everyone. "Today's your lucky day, my man." Kraft's much smaller but he presses himself as close as he can without hurting Pieciul. "So . . . what should we talk about?" Kraft asks, trying to keep his tone upbeat.

* * *

After two hours of his friends telling him goofy stories about ski wipeouts, Pieciul wonders how much longer they can keep it up. One tale from Kraft is interrupted by Tsirku's sudden growl aimed in the direction of a nearby grove of spruce trees. *Is that the bear? Has she come back? I can't fight back at all . . .*

Pieciul feels his heartbeat quicken, and his body goes tense with fear. Moskowitz and Kraft start banging on metal avalanche shovels and shouting as loud as they can. Pieciul shouts too, as best he can, anxiously scanning the darkness of the grove for any sign of movement.

The trio's din persists for minutes, until Tsirku starts to relax. Pieciul has no idea if the bear was really there or if something else spooked the pooch. He only knows this waiting is mentally and physically exhausting.

"How much longer?" he asks Moskowitz, who pulls out the inReach and presses a few buttons.

"An hour, maybe less," Moskowitz says.

Pieciul musters a grin. "Oh, that's easy," he pants. "We can do an hour."

Kraft starts telling another story and Pieciul is trying to follow along, but his mind drifts to the bear who attacked him. *They can't kill her. I came here for the wilderness. Bears are part of that deal. She could've killed me, but she didn't.*

"It's not her fault."

Pieciul realizes he said that aloud. Kraft stops his story. "Whose fault, Bart?"

"The bear. She wasn't trying to hurt me. She wanted me away from her cub. That's all. Don't let them hurt her."

The next twenty-five minutes pass slowly, but soon enough there's a faint mechanical whir of helicopter blades. As it grows louder, Moskowitz and Kraft shovel out a landing zone for the rescuer, who descends quickly with a medevac basket.

The rescuer kneels, shouting over the rotor noise: "I hear you danced with a bear."

"Just a little tango," Pieciul replies.

After he's cinched into the litter, Pieciul stares at his two friends, his saviors. His eyes well up as Kraft rubs his shoulder. Moskowitz flashes a grin and a thumbs-up, adding, "You got this."

"Thank you," Pieciul says to them before the basket is hoisted into the air. "For everything."

*　*　*

Bart Pieciul was taken to a Juneau trauma center, where he was whisked into surgery. The results were surprisingly positive. He regained full use of his hand, and a GoFundMe, set up by Moskowitz and Kraft, fully covered his medical bills. He was enjoying the backcountry within months of the attack. He got his wish too: Wildlife officials decided not to hunt down the protective mother bear, agreeing with Pieciul that she was rightfully defending her den.

HOW TO SURVIVE A BEAR ATTACK

KNOW YOUR BEARS. Different areas are home to different bear types, and understanding whether you're in brown/grizzly bear territory or black bear territory is crucial to knowing how to respond to an attack.

CARRY THE RIGHT GEAR. Moskowitz's backpack had all the essential first-aid gear needed to keep Pieciul alive, including an inReach satellite phone. These transmit SOS signals but also can send text messages to other inReach devices, so you have multiple means of sending a distress call. Carrying bear spray—stronger pepper spray with a reach of forty feet—is a good idea if you'll be anywhere near bear territory.

UNDERSTAND BLUFF CHARGES. These are highly common moves bears use to scare off a potential threat. In a bluff charge, the bear may rear up on its hind legs or puff itself up to look more threatening. Its ears will be up and forward, facing you. It will make several startling leaps toward you, but before it reaches you, it'll stop or dart off to the side. There will also be some growling or other vocal notes from the bear.

DON'T RUN. If a bear is bluff charging, running away may initiate a chase and attack. Instead, back up slowly while facing the bear. Make yourself as big as possible, waving your arms over your head, and speak clearly and slowly to the bear in a loud voice, to make it clear that you're a human.

KNOW BEAR ATTACK WARNING SIGNS. These include yawning, clicking teeth, pounding front paws into the ground, and ears pinned back against the side of its head. If you see these signs, the bear is stressed and could charge you for real at any moment.

IF IT'S BROWN, LIE DOWN. When dealing with a charging grizzly, play dead. Go completely limp, but cover your head and face with your hands. If you're able, lie on your stomach and use your backpack or other gear as a shield to protect your spine and back. No matter what happens, try to resist making any noise.

IF IT'S BLACK, FIGHT BACK. If a black bear charges you, your best hope is to fight violently. Playing dead doesn't work with black bears—but fortunately they're smaller than grizzlies, so while they can still be deadly, you at least have a chance. Hit it in the head, face, and eyes with everything you have, including rocks, tree branches, or other backcountry gear.

WAIT BEFORE MOVING. If you've survived an attack, let several minutes pass before getting up. Bears want to ensure you're not a threat, so if you pop up quickly, it may return and maul you again. Listen very carefully to the bear leaving the area and then slowly get up and find a safe place to address your wounds.

TREAT WOUNDS WITH MARCH. For anyone injured, use the MARCH approach. It addresses critical issues in the correct order: massive hemorrhage, airway, respiration, circulation, hypothermia. "This is how combat medics approach battlefield wounds," says Sachdeva. "Your biggest threat of death is massive hemorrhaging from a wound, so you want to control this first. Use a tourniquet or quick-clotting powders," says Sachdeva. "Next, make sure the victim is breathing; check the airway to make sure it's not obstructed. Respiration and circulation are combined; if the victim isn't breathing, you need to start CPR to restore breathing and circulate the blood. All of this is trying to get oxygen to the tissue." Last, treat hypothermia by keeping the victim warm with layers and by starting a fire.

ATTACKED BY A CROCODILE

8

NOVEMBER 10. AFTERNOON.

As Craig Dickmann casts his fishing line into the Coral Sea, he knows he made the right choice. He's standing on a rock platform, a ledge of stone jutting out into the sea, here at Captain Billy Landing, a remote area on Australia's northeast Cape York Peninsula.

Earlier, the fifty-four-year-old wildlife ranger from Queensland, Australia, had been deciding between mowing his lawn or going fishing. But here, with the azure waters lapping below, he wonders why he'd even hesitated. The conditions today are nearly perfect for fishing—a warm breeze, a sandy beach, and blissful solitude. The only detractor: It's overcast, and Dickmann thinks a storm may be coming.

The fish aren't biting. He throws a few more casts into the water and then glances west. Dark storm clouds are closing in on him. He sighs and starts to reel in his line. Time to go home.

Then out of the corner of his eye, he spots a flash of brown and tan. He knows immediately what it is—a saltwater crocodile. As a ranger, Dickmann is familiar with the dangers of Australian wildlife. Before he started fishing, he'd checked the area around the rock platform to make sure the ground was free of crocodile tracks. He hadn't seen any, so this croc has come out of nowhere.

It's coming not from the water but from the rock platform behind him. Before he even has time to register the danger he's in, the croc has closed the distance between them and is lunging at his leg.

It's a big one, about nine feet long. A blur of teeth glistens in the sunlight, then there's a loud crack as the croc goes for a bite. Dickmann knows this horrifying sound; a croc's jaws are so fast that, when they close, the air forced out causes a loud popping or snapping noise.

The first bite misses. The second doesn't. The croc latches on to Dickmann's left thigh. The pressure is otherworldly. Saltwater crocodiles have one of the strongest bites in the animal world, and Dickmann feels every ounce of that crushing force as the three-inch teeth dig into the meat of his thigh. The croc's jaws clamp tighter, and Dickmann falls to the ground.

> **Bites alone can snap your bones. A crocodile bite can apply five thousand pounds of pressure per square inch. By comparison, humans bite at a mere one hundred pounds per square inch. Worse, if bitten, you're dealing with crushed muscle, arteries, and blood vessels. "Crushed muscle releases cellular contents and toxins into surrounding tissue," says Matt Cummins, a board-certified emergency physician. "Those areas swell, compressing the blood vessels. Without blood-carrying oxygen, the tissue dies. You have six hours before you'll lose that limb." Your kidneys will try to clean up the toxic muscle cells, but this often leads to kidney injury and failure.**

Dickmann thrashes but the croc won't let go of him. Blood is coursing from his leg. He feels the beast dragging him in the direction of the surf. *It wants to finish me off in the water*, he realizes.

Dickmann jams his thumb into the animal's eye socket, the only weak point in the beast's otherwise armored body. It only aggravates the animal. The croc retaliates by initiating a death roll, a dizzying spin used to disorient and stun prey, usually done in the water. This 360-degree sideways roll helps rip prey apart. The crocodile hasn't found any bone yet, only soft tissue and muscle. But its teeth keep digging into Dickmann's skin—piercing it, shredding it, tearing it.

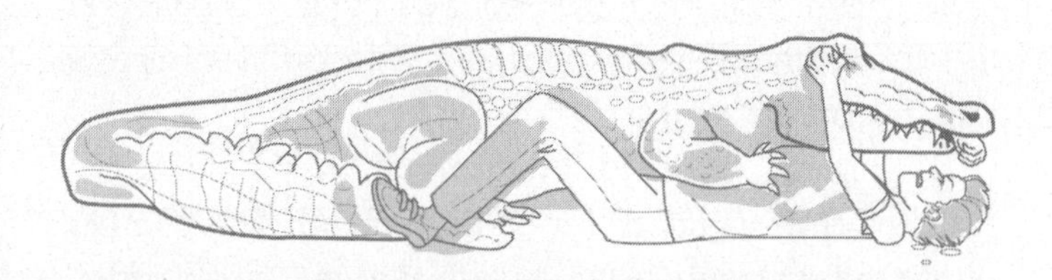

> **When attacking, crocodiles stun and then overwhelm you. The first bite is usually followed by a death roll, as Dickmann endured. The death roll aims to thrash its prey against something, to aid in dismemberment. Underwater, crocs will roll prey against rocks or the bottom of the seabed. On land, the death roll is equally effective. The faster they can flip you, the better the chances of tearing you into smaller pieces to be eaten.**

Death rolls can snap an arm or leg instantly. Any bone break is problematic, but when twisting, those broken ends cause worse damage. "If a croc is spinning you, broken bones can have jagged edges, which act like sharp knives," says Cummins. "You will have massive tissue and vascular damage. And bleed a lot more."

The two struggle on the ground of the rock ledge, Dickmann on his back, the crocodile's four hundred pounds on top of him. Its jaws fly open as it climbs up his body. Dickmann's hands dart up to protect his face. Another sickening snap. His right wrist is now inside the croc's mouth.

He whacks at the snout with his left hand. He doesn't have a clear shot at the eyes, but he's trying. His right arm is locked in the croc's jaws as the animal continues dragging him toward the sea.

Dickmann's thumb finally connects with the croc's eyeball. Another mighty thrust of his thumb triggers another death roll. This spin peels the skin and some muscle from Dickmann's right hand. His flesh is completely removed. What remains of his hand is a bloody mass of pulp and bone.

If your skin is completely ripped off your hand, that's called degloving. Skin and tissue are removed, exposing your muscles, nerves, and bones underneath. It's an extremely painful injury, because you're exposing nerve endings that should never feel air. "Exposed muscle dries very quickly too," says Cummins. Keep any exposed muscle and tissue wet, ideally

with clean water, but if the only thing available is scummy pond water, use it, says Cummins. "You'll get antibiotics later; dry tissue will die."

The croc is on top of him again. Dickmann doesn't have much fight left in him. If those teeth sink into his neck or an artery on his arm or leg, he's done for. With his bloodied wrist still in the croc's mouth, the croc shakes its head, as if it knows victory is imminent.

With his left hand, Dickmann's fingers search the croc's bony, scaly skin. His thumb finds an eye again: *Push it into the brain. Give it everything you've got.*

Dickmann's thumb plunges so deep into the crocodile's eye socket that he can feel bone. It works; the startled croc releases his hand. Dickmann scrambles from underneath the hulking reptile and swings a leg over its back, straddling it now. He presses his uninjured hand down on the closed jaws. He leans all his weight into it. Crocodiles evolved to clamp down hard on their prey, but the muscles that open their jaws are relatively weak. Dickmann knows this and begins to feel a shift in this battle. If he can just keep pressure here.

An eerie quiet descends over the scene. Dickmann can hear the water lapping onto the shore, the sound of his panting as he leans down upon the creature's snout. Blood soaks them both. What now?

The croc acts first. It starts to scuttle toward the water, effortlessly carrying Dickmann on its back. Keeping pressure on the snout with his good hand, Dickmann hurries to his feet. He stands beside the croc, bending over. In this awkward position, Dickmann shuffles along with the croc as they near the water's edge. He wishes he could just hurl the whole crocodile into the water, but it's too big. Several bones in his right hand are broken, but he can move his fingers. He grips the snout with both hands, clamping it closed while pulling upward. Adrenaline surges and, somehow, he's able to get the croc's head off the ground.

> **When adrenaline surges, you're able to tap into superhuman strength. Your heart and lungs work faster, contracting your blood vessels to direct more oxygenated blood to your muscles. The muscles also get extra boosts of glucose, giving you more energy to use. The increase of blood flow is also removing unwanted elements from your muscles, including carbon dioxide, which means you can work harder for longer without cramping. Because of all this, your muscles get a sizable power boost. This is how Dickmann was able to wrestle with the crocodile and lift its snout.**

Dickmann can't muscle the croc's head much higher than his waist. The beast is too heavy and it's starting to thrash. Swiveling, he throws the snout as far as he can. There's a splash, and the croc slips beneath the surface. A final swish of the tail, and it disappears.

Dickmann's next thought is an odd one. *Get your shoes.* They'd fallen off during the struggle. He looks around the water, searching for, of all things, his Crocs. His brain catches itself a moment later. *Idiot. Get the hell out of here.*

> **Why did Dickmann immediately look for his shoes after the attack? "That's shock," says Anthony Giovanone, a doctor of osteopathy and a psychiatrist. "The brain dissociates as a response to trauma to protect itself from that intense emotion temporarily. What just happened isn't even coded in his brain yet. His brain is still focused on finding lost shoes."**

He stumbles to his car, assessing his wounds while moving. He thinks his mangled hand looks like an anatomy textbook, with layers of muscles, nerves, and bones all visible. It's a horrendous sight. There are puncture wounds on his calf; he hadn't even felt that bite. The thigh bite is very deep, and blood is pumping from there. Dickmann knows this is a grave situation.

In this remote part of Australia, there's no cell phone service. There's no one else around. And it's an hour's drive to the nearest help—Heathlands Ranger Station, where Dickmann works. That's where he will have to head. But first he needs to stem the bleeding. Fumbling with a first-aid kit he keeps in the car, he manages to pack some gauze onto his degloved hand and wrap it up. The gauze turns red instantly, but there's no more to use.

Toward the end of the drive, Dickmann begins to feel dizzy and faint. Blood drips everywhere, soaking his outfit and the car's seat. The steering wheel is slick from his bleeding hand too. But he reaches the ranger station. He struggles to open the car door and staggers to the station house. His limbs aren't cooperating. He almost falls several times trying to reach the station door. Pushing into the building, he faintly calls out for help. Silence. No one's here. He slumps in a chair and radios a colleague elsewhere on the station's base to come help, hoping they arrive quickly.

> "If you survive an attack, you could be bleeding heavily. We can lose up to 750 mL of blood—equivalent to a bottle of wine—without showing any symptoms, says Cummins. This is about 15 percent of your total blood volume. If you lose up to 1.5 liters, or 30 percent of your blood, your heart and respiratory rates will increase. "The heart has to work faster to pump blood around you," says Cummins. "You won't feel dizzy or pass out until you lose two liters of blood. Then your blood pressure drops, you have an altered mental state, and you'll be almost in and out of consciousness." Lose more than 40 percent of your blood, and that's impending death, per Cummins. If a wound is severe enough, you can die from blood loss in as little as two minutes."

It's six o'clock, two hours since the attack, when his colleague appears. The nearest ambulance is three hours away, and it's eleven hours by road to the nearest hospital, in Cairns. That's too far. He needs to be airlifted out. Now. Except it's too dark to land at the tiny airstrip a few minutes from Heathlands Ranger Station.

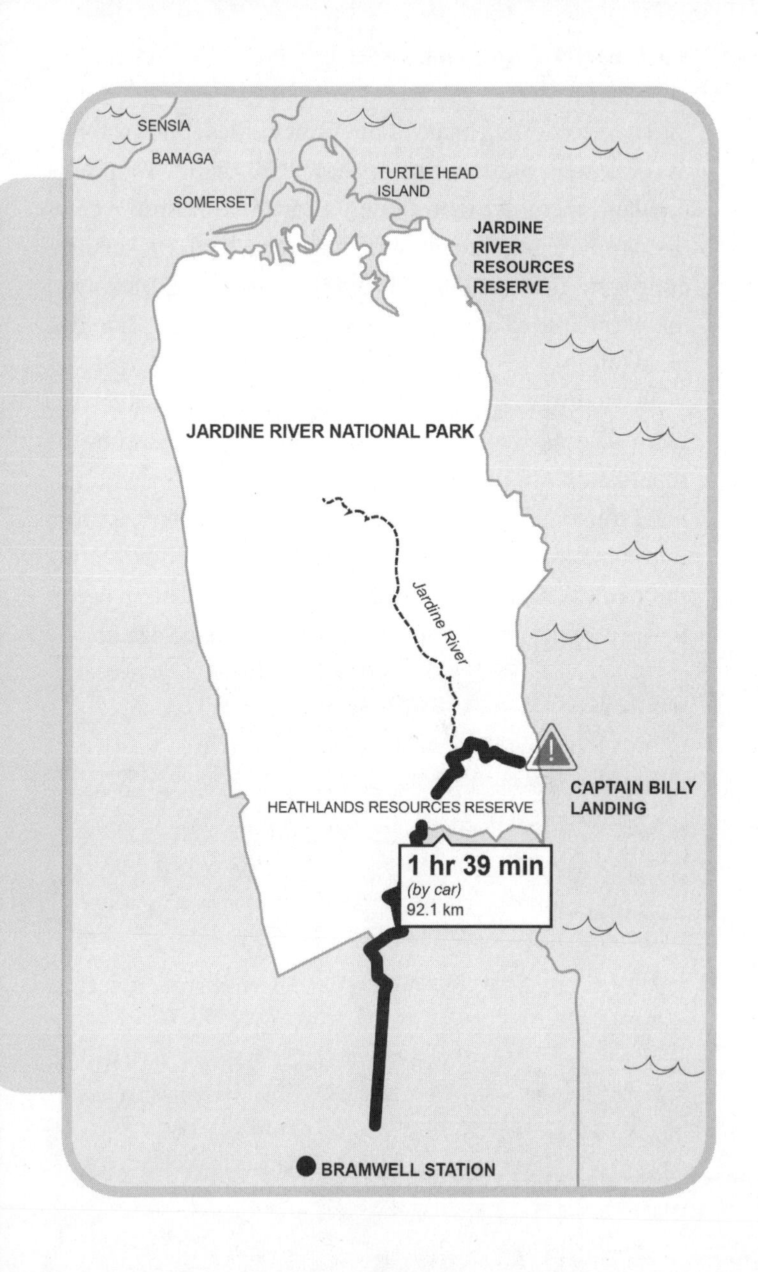
SENSIA
BAMAGA
SOMERSET
TURTLE HEAD ISLAND
JARDINE RIVER RESOURCES RESERVE
JARDINE RIVER NATIONAL PARK
Jardine River
CAPTAIN BILLY LANDING
HEATHLANDS RESOURCES RESERVE
1 hr 39 min
(by car)
92.1 km
BRAMWELL STATION

There's another, bigger airstrip fifty miles south, though, at Bramwell Station, which has solar lights so planes can land there at night. It's on a cattle farm, so the cows need to be cleared to allow the plane to land. Dickmann's colleague alerts Bramwell and then patches Dickmann up with limited supplies from a medical kit. The two head off.

By now the analgesic effect of all the adrenaline has worn off, and the pain is excruciating. Every bump in the road sends a fresh jolt of agony through Dickmann. His hand throbs. He can feel his heartbeat pulsing weakly from the puncture in his thigh. He wonders if he'll die during the endless ninety-minute drive. And even if he makes it to the hospital, will he ever have use of his hand again?

> Adrenaline is also responsible for Dickmann not feeling pain from his massive leg wounds or broken hand during his fight with the crocodile. "Adrenaline and the body's natural endorphins that kick in during situations like this alter pain perception," says Beth Palmisano, a medical doctor who specializes in pain management. "It doesn't alter the pain itself; the chemicals instead affect the brain's central processing of pain. These downregulate pain signals, and when our spinal cord detects these neurotransmitters flowing, the cortex of our brain decides we must get away from danger. It'll supersede our pain perception so we can focus on survival, not the hand that just got degloved."

The plane is waiting when they reach Bramwell. Medical staff hurry him onboard. He manages a few jokes on the flight because, for the first time in four hours, he knows he's going to live.

* * *

The crocodile's first bite narrowly missed Craig Dickmann's femoral artery. He suffered extensive tissue damage, but surgeons repaired it. Surgeons were also able to reconstruct his degloved hand, and he has no permanent issues after the attack. The crocodile was later found and euthanized.

HOW TO SURVIVE A CROCODILE ATTACK

Take a page from Craig Dickmann's book. When entering a known crocodile area, first check for croc tracks, keep dry land behind you—crocs are less likely to attack on land—and stand back from the edge of the water or any brush. If the worst occurs and you're staring at a crocodile, here's how to escape.

ON LAND, BACK AWAY SLOWLY. Crocs can reach top speeds of 10 mph on land, but they tire easily. They'd prefer their prey to be in the water. Ease backward, keeping the croc in your line of sight.

IF CHASED, RUN STRAIGHT. It's a myth that running in a zigzag pattern is the best option. Sprint in a straight line and keep your eyes ahead. If you look back, there's a greater chance you'll fall. Run away from the water.

DON'T SPLASH. If you're in the water and the croc approaches, avoid sudden jerky movements. They'll draw unwanted attention. Instead, calmly (or as close to calm as you can) get to the water's edge and continue backing up.

IF YOU'RE BITTEN AND RELEASED, RUN AWAY. If at all possible, run after an initial bite. Fast. If the croc didn't hang on to you, it's likely a defensive or territorial bite and the croc isn't interested in eating you.

IF BITTEN AND HELD, FIGHT. Gouge, stab, punch; do whatever you can to aggressively let it know you're not going down easily. Go for the eye, as Dickmann did—it's the most vulnerable part. Use extreme force when pushing. As Dickmann found, it can

take several tries before succeeding.

AVOID BEING PULLED INTO WATER. Crocs have serious advantages in the water. Their death rolls can be more overwhelming underwater. They can hold their breath for long periods and use their immense tails as weapons too. Do whatever you can to stay on land. "If a croc takes you underwater, you're not coming back," says Cummins.

ROLL WITH THE CROC. If caught in a death roll, move with the crocodile. It'll reduce the strain on your body and may prevent your limbs from being torn from your sockets. Don't stop fighting, though. Continue to go after the eyes.

TREAT BLEEDING IMMEDIATELY. "If you're squirting blood, that's the biggest issue," says Cummins, adding that even small arterial bleeds will squirt. "Squirting implies higher pressure. Any blood coming from the heart to the organs can squirt." To address this, "use a tourniquet—a belt, a shirt, whatever you have—and tie it six inches above the wound. Use a stick to twist it incredibly tight. It will hurt a lot, but you have to stop the blood flow."

If blood is oozing, that's a low-pressure, or negative, bleed—and it's from blood coming back to the heart. These can be easier to treat, and for wounds that are oozing, Cummins recommends pressure bandages but says a T-shirt with a backpack strap around it can work. Both squirting and oozing blood will cause a massive loss over time, though it'll happen faster if you're squirting. If you're applying a tourniquet, get to medical help within six hours, or else you risk amputation.

IF DEGLOVED, KEEP THE AREA MOIST. First, stop any large bleeding with pressure, says Cummins. Rinse the wound with clean water, clearing away any dirt. If you can place the skin back on, do so. Wrap the wound tightly and keep wetting the bandage with water. "You don't want anything drying out before you reach the hospital," says Cummins. Dickmann did the best he could when applying gauze before driving for help, and his profuse bleeding likely kept his tissue from drying out.

GIVE UP A LIMB. If you're caught in a croc's mouth and people are trying to pull you to safety, let them. "A croc can absolutely rip your arm off," says Cummins, adding that this is preferable to being dragged underwater. Should this happen, pack and wrap the wound very tightly. Cummins recommends carrying quick-clotting powders, medical chemicals that cause a heating reaction that cauterizes open wounds. "It's what soldiers use in battle," Cummins says.

POISONED BY A RATTLESNAKE

MAY 27. MORNING.

Jeremy Sutcliffe takes one hand off the handle of his lawn mower and wipes the sweat from his brow. It's only ten thirty in the morning, but the South Texas humidity is so thick it's like breathing through a wet blanket. It's shaping up to be a scorcher. Forty-year-old Sutcliffe and his wife, Jennifer, are hosting a cookout this afternoon, so they're both tidying up their one-acre property near Lake Corpus Christi. Jennifer is in the backyard pulling weeds from the garden.

Above the growl of the lawn mower, he hears a scream.

"Snake! SNAKE! HELP!"

It's Jennifer. Jeremy kills the mower engine, grabs a shovel, and races toward the backyard. There he sees Jennifer, her back pressed against the side of the house, her face white, frozen with fear. She's pointing into the garden, where she'd been weeding.

> **Beyond fight or flight, you can freeze. When Jennifer first froze, that's normal. "When a threat is detected, you stop to orient," says Anthony Giovanone, a doctor of osteopathy and a psychiatrist. However, when Jennifer later sees Jeremy fighting the snake, her medical training activates her prefrontal cortex, enabling higher-level thinking. "The amygdala can handle quick, basic survival action. But problem-solving**

thinking—get the car keys, call 911—wins out and you're able to get moving." Slow, deep breaths can give the prefrontal cortex a chance to come back online so the amygdala isn't the only thing running the show, says Giovanone.

"Is it a rat snake?" Jeremy asks. He's a serious outdoorsman who camps and fishes often, and he knows these harmless reptiles sometimes crawl around the property.

But Jennifer doesn't respond; she just keeps pointing. His gaze follows her shaky hand, which directs his attention to a thicket of weeds.

A dusty brown triangular snake head peers back at him, tongue flicking. He hears a hollow rattling sound. Jeremy knows precisely what he's looking at: a Western diamondback rattlesnake.

Western diamondbacks are among the most lethal of snakes, and this one's huge, about four feet long. It's agitated and has lifted its head about a foot off the ground into a striking position. Using the shovel as a shield, Jeremy steps cautiously toward the rattler. He stretches the shovel out toward the snake, trying to scoop it up. The snake lunges at the shovel blade with a menacing hiss. *This isn't working*, he thinks. *Even if I can get it onto the shovel, what will I do with it?*

"Western diamondback rattlesnake venom is exceptionally nasty and lethal. If untreated, your chances of dying from the venom are between 10 and 20 percent. Once bitten, the viper's toxins instantly start killing you."

He flips the shovel around, takes a step closer, and smashes the sharp edge down on the snake's body, a few inches behind the head. The blow works; the snake is decapitated. Its headless body writhes on the ground, a bloody, scaly tangle that's repulsive to watch.

"It's okay," he says to Jennifer. She's freaked out from this encounter, and he can tell. "You go in and catch your breath."

This story is an interesting example of the body's sympathetic nervous system response, says Giovanone.

Fight or flight kicks in as an early response to stress, such as seeing a snake. How? "In our primitive visual center, we see things in black and white," says Giovanone. "That image gets to the amygdala—your brain's emotional and fear processor—the fastest, in a split second. Then higher-order perception kicks in. The black-and-white image could look like a snake. But when higher perception comes online, you realize it's a stick. The amygdala is determining whether this is a threat or to chill out."

If threatened, the hypothalamus snaps into action. "This is where your hormones join the party," says Giovanone. "Hormones issue orders, like *Give more oxygenated blood to the muscles so you can run*, or *Dilate the pupils to let more light in.*" This sympathetic system response happens in milliseconds. Your body moves from rest and digest into a state of readiness, pulling blood from your gut and skin and parking it in the muscles, where it's likely most needed. This is why people look pale during tense moments, says Giovanone.

When he hears the back door slam shut, Jeremy works the blade of the shovel under the snake's body and lifts it off the ground. It's still squirming, but he dumps it into an outdoor trash bin. He closes the lid and can hear its death throes as it bangs against the sides of the bin. He goes inside to check on Jennifer, who still looks pale and shaken. But her breathing has slowed down at least.

"Can you just get rid of the head?" she asks. "I want to let the dogs out."

"Absolutely," he says. "Give me a second."

He steps back outside. The snake's head is resting on a paving stone, in a pool of blood. It's not moving. There's a big stick on the ground just beyond the head. *Perfect for flicking it away*, he thinks. He reaches down, his fingers passing inches from the snake's snout.

Suddenly the snake's jaws open. This thing isn't dead.

Before Jeremy can yank his hand back, the snake strikes. Long fangs sink into Jeremy's fingers on his right hand, so deep they hit bone. The rattler releases a huge shot of venom. The pain is instant and excruciating, like his hand is being crushed in an enormous vise. Jeremy lets out an agonized scream.

> **Snakebites are extremely painful due to the irritation of the venom going into the skin. "Snake toxins have inflammatory factors," says Dr. Deepak Sachdeva, an emergency room physician. "Ever get an injection of a local anesthetic? That stuff burns. When toxins hit fast, it burns like crazy." That immense smashing sensation is pressure from your tissue swelling. "Your body is reacting to the inflammation in the area, but also the venom," Sachdeva says.**

"It bit me!" he yells. He violently shakes his right hand. The snake refuses to release, its fangs still buried in his flesh. Blood drips from the severed end. Searing pain moves up his arm as venom spreads throughout Jeremy's body. His hand is already starting to swell. It feels like his hand is being hit with a sledgehammer, over and over.

> **Venom can spread quickly. "First you get local swelling, near the area of the bite; then you get the systemic response, when it starts dropping your blood pressure, affecting breathing and neurologic functions," says Sachdeva.**

The snake will not budge.

He shrieks again and claws at the rattler's head with his free hand. He manages to get a few fingers from his left hand into the snake's mouth. Tugging hard, he frees one of the two fangs, but as it clears Jeremy's skin the reptile's jaws snap shut. The freed fang sinks into the ring finger of his right hand. A fresh dose of venom is released. This time the feeling of pressure is so immense that it feels like his hand may burst like a balloon. He needs to get this thing off before he passes out.

* * *

Rattlesnakes are ectothermic creatures, meaning they need external heat sources to stay alive. Ectothermic beings have a lower need for oxygen in their organs and brains to stay alive. That's why even after being decapitated, a snake's head and body can continue to live. There are reports of severed snake heads biting up to forty minutes after being separated from their bodies.

Most rattlesnake bites are defensive, and little venom is released, if any at all. Producing venom takes a lot of energy and snakes are judicious in how much they inject. However, when a snake is dying, it has nothing to lose. The severed rattler head releases all of its venom into Jeremy Sutfcliffe's hand because it's threatened and in immense pain. Jeremy's struggle with the snake means that his blood is pumping faster, distributing the venom throughout his body quicker.

> **Rattlesnake venom is a mix of neurotoxins and hemotoxins; it kills your blood cells and tissue, as well as your nerves. It attacks your circulatory system, affecting your blood's ability to clot, and causes internal bleeding. It also destroys your nerves, leading to immense pain and swelling of tissue.**

At the sound of her husband's screams, Jennifer throws open the back door and rushes toward the garden. What she sees is horrifying. Jeremy spins around madly, the decapitated viper head hanging grotesquely from his hand. Jennifer is a nurse, and her training instantly kicks in. He needs medical help and antivenin. Immediately. Without a word, she runs back inside for the car keys, while Jeremy keeps fighting to rid himself of the zombie rattler head.

She grabs the keys off the kitchen counter, then dashes back outside just as Jeremy finally manages to pry the head off, flinging it far away. "Get in the car," she yells. She's already dialing 911.

Around the hospitals where Jennifer has worked, she's heard the conventional wisdom concerning snakebites: Time is tissue. Venom kills cells and tissue instantly; the longer a body goes without antivenin, the more severe the damage. If she doesn't get Jeremy help soon, he could lose his arm. Or he could die.

* * *

Jennifer floors the gas pedal as the car races through her neighborhood. She's on the phone with a 911 dispatcher, who has bad news: The nearest hospital doesn't have any antivenin. The closest ER that does is an hour away. She glances over at Jeremy, who's in the passenger seat. He's starting to lose consciousness.

He blinks rapidly. "I can't see," he moans. "It's just black." It's been ten minutes since the bite, but Jennifer knows he may not survive the drive. Jennifer reaches over and shakes him with her free hand. He's got to stay awake. She feels him start to convulse. It's a seizure. The venom is strangling his circulatory system, shutting it down.

> **Seizures are rare, seen only after high doses of snake venom, as Jeremy Sutcliffe experienced. They happen as a result of the toxins interfering with your nerves and nervous system. "It's a circulatory collapse either from your dropping blood pressure or from the snake's neurotoxins," Sachdeva says. You can develop confusion or lose consciousness.**

Fighting to keep the panic out of her voice, she says, "You're going to be okay." But she's terrified she could be wrong.

Jeremy snaps back to alertness, but it's brief and he passes out again in seconds. The 911 dispatcher is still on the line and directs Jennifer to pull into a nearby parking lot; an ambulance will meet them there. She finds the parking lot, and for fifteen agonizing minutes they wait for the ambulance to arrive. Jeremy drifts in and out of consciousness. When he's awake, he blabbers incoherently. But then he locks eyes with his wife.

"If I die, I love you," he whispers.

She squeezes his shoulder and tells him help is nearly there. Another seizure starts as the ambulance arrives. Paramedics strap him into a stretcher and load him in the back. They start stabilizing him, trying to get his blood pressure steady. Within a minute, the ambulance is racing toward the hospital, Jennifer speeding behind it. After a few miles, the ambulance pulls over into a church parking lot. *What are they doing?* Jennifer thinks. *This is no time to stop.*

Screeching to a stop, Jennifer hops out. A medic throws the ambulance door open. He explains that Jeremy's blood pressure has plummeted and they can't increase it. He needs an emergency room, and the quickest way now is via helicopter. Jennifer can hear the rotors now. Within seconds, the chopper has set down in the parking lot. Jeremy is loaded and in the air in seconds.

> **Your blood pressure drops because the snake's toxin is decreasing the ability of your blood vessels to properly constrict. "If your blood vessels aren't constricting properly," says Sachdeva, "there's less pressure. When your blood pressure dips, there's less blood flowing to your vital organs, which causes a slew of other issues."**

* * *

At the hospital, a team of seven doctors has been working feverishly on Jeremy for more than five hours, getting him fluids and trying to get his blood to clot.

> Internal bleeding can start in minutes. As it kills your blood cells, a rattler's poison can both cause clots and increase bleeding, says Sachdeva. "The toxins can disrupt the body's clotting cascade—a system of proteins that control your blood's ability to clot without clotting too much. If your blood can't clot, it'll affect the whole body," Sachdeva says. "It's like a system of hoses that springs a million leaks. You treat the clotting problem—with antivenin—and hope the leaks stop." This is a primary reason why Jeremy's blood pressure couldn't be stabilized.

They had to insert IV bags into blood-pressure cuffs and inflate them in a bid to squeeze liquids into him as fast as possible. This was a method Jennifer had never seen before. But doctors couldn't stabilize his blood pressure, and his organs were failing.

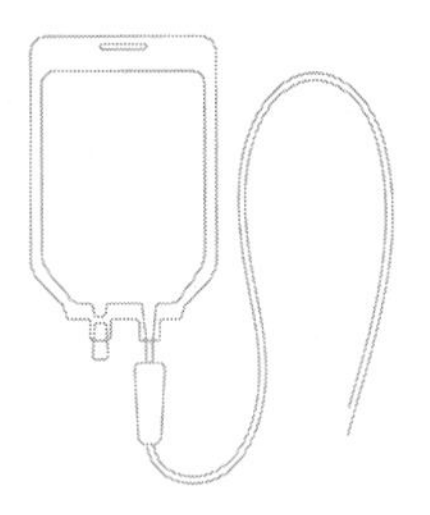

> As your body bleeds internally, you can enter shock. "Shock is the body's inability to get circulatory system pressure to deliver blood and oxygen to tissues," Sachdeva says. "Jeremy likely had both distributive shock, which is when your blood vessels aren't constricting enough to maintain blood pressure; and hemorrhagic shock, when there's not enough blood volume left to be distributed." The longer you're in shock, the more tissue dies. You also risk infection. Regardless of the type of shock, the symptoms are the same: a rapid heart rate, low blood pressure, cool and clammy skin, and mental status changes like confusion or even a coma.

Most snakebite victims need between two and four rounds of antivenin. Jeremy received twenty-six doses, with some needing to be flown in from other hospitals.

> Antivenin is composed of antibodies that bind to the venom components to neutralize their toxic effects. It's made by introducing small, harmless amounts of venom to animals, such as horses, sheep, or dogs, and then harvesting the antibodies from those animals. (These animals aren't harmed and have robust immune systems.) Normally, two to four doses of antivenin are effective in preventing permanent damage.
>
> But at high doses of antivenin, as Jeremy received, anaphylaxis can occur. This is a severe form of allergic reaction, which results in shock-like symptoms—a rapid heart rate or an altered state of consciousness. These symptoms can mimic venom symptoms, and doctors may not know the culprit, as

Sachdeva admits. "You keep doing the antivenin treatment," he says. "Then treat any allergic reactions with steroids and antihistamines."

The rattlesnake venom caused extensive damage to Jeremy's kidneys and the nerves within his abdomen. Dying tissue had caused his kidneys to completely shut down, and he was put on dialysis. He was placed into a medically induced coma and put on a ventilator. Beyond this, little could be done. Three separate times, doctors told Jennifer that Jeremy was probably going to die.

Four days later Jeremy woke up.

Doctors tried hyperbaric chambers and skin grafts to save his two fingers, but the necrosis kept spreading and the bitten fingers were ultimately amputated. Gallstones, kidney stones, and unbearable abdominal pain wracked his body for weeks. The nerve damage to his stomach was irreparable and he now suffers from megacolon, an abnormal dilation of the colon. Nerve pain all over his body lingers to this day.

Damage can be permanent. When skin or soft tissue dies, there's a small window to revive it. If the toxins prevent that, then the dead tissue needs to be removed. If that's on a finger or toe, amputation can be necessary. Nerve damage may not be reversible either. "Nerves don't regenerate well," says Sachdeva. "If they do, it's extremely slow. It's rare to have a full recovery."

NO EYELIDS, BUT EYES PROTECTED
WITH SHEDDABLE CLEAR SKIN

HOLLOW FANGS LIE FOLDED BACK IN THE MOUTH,
UNTIL THEY NEED TO INJECT POISON INTO VICTIM

THICK BODY WITH
DIAMOND-SHAPED
SPOTS

**WESTERN DIAMONDBACK
RATTLESNAKE**

HOW TO SURVIVE A RATTLESNAKE ATTACK

Avoid rattlesnakes whenever possible—wear long pants, tall boots, gloves, and other protective clothing when moving around in rattlesnake territory. But if the worst happens, here's how to handle a snakebite and the subsequent dose of venom.

GET AWAY FROM THE SNAKE. If you're bitten, move away—fast. You don't want it coming back for another round.

NOTE THE SNAKE'S APPEARANCE. Don't try to catch it, but remember its color, shape, and size. Identifying the type of snake that bit you is vital to getting the correct antivenin at the hospital.

ACT IMMEDIATELY. You're racing against toxins that will only cause more damage the longer they spread. The ideal time to start antivenin treatments is within a half hour of the bite. Call 911 and get help.

DON'T SUCK THE WOUND. Don't try to suck the venom out; studies show it doesn't help at all and can only introduce bacteria from your mouth, making infection more likely. You can wash or clean the wound, Sachdeva says. It's a myth that cutting the wound larger helps; this increases your chances of infection, so don't do it.

LET IT BLEED. Light bleeding will bring some venom with it. Never apply a tourniquet; any constriction increases the risk of amputation. If it's bleeding heavily, Sachdeva recommends applying direct pressure and covering the wound with a clean cloth.

REMOVE TIGHT CLOTHES AND JEWELRY. Before your skin swells, get rid of anything constricting, including rings, necklaces, and watches. Quickly discard any tight clothing too, or cut slits in it to allow it to handle impending swelling.

DON'T ELEVATE THE BITE. Keep the wound and bite area below your heart. Elevating it may help reduce swelling, per Sachdeva, but it can also cause the venom to spread faster.

KEEP STILL. Moving helps the venom circulate faster, so avoid motion. Try to be calm and slow your breathing too. This can help slow the toxin's spread.

CHASED BY A WOLF

10

JULY 6. AFTERNOON.

William "Mac" Hollan pushes hard on the pedals of his road bike. It's midafternoon and he's zipping along the Alaska Highway, sixty miles west of Watson Lake in Yukon, Canada. He's halfway through a six-week, 2,750-mile trek from Sandpoint, Idaho, to Prudhoe Bay, Alaska. Hollan is riding for charity, fundraising to benefit the elementary school in Sandpoint where he teaches.

A hardcore long-distance adventurer, the thirty-five-year-old Hollan has done several lengthy bike rides for charity, including one 4,200-mile cross-country trek for a children's hospital a few years back. Joining him on this ride are two close friends, Gabe Dawson and Jordan Achili. The trio are camping at night and covering eighty miles per day, six days a week.

But Hollan's buddies are currently lagging at least a half mile behind him on this stretch of road, around a bend. A few cars pass, but the road isn't terribly busy. Towering pine trees line a small incline to his right; snowcapped Canadian mountains rise in the distance. It's a view that can't be beat.

A rustling noise from behind startles him. *Must be Gabe or Jordan catching up?* He turns back over his shoulder, but instead of his friends, he sees . . . a dog? It's a big one too—at least one hundred pounds. And it's running up the shoulder of the road, a black and tan blur. Hollan is confused for a second. What's a dog doing out here in the middle of nowhere? But Hollan's confusion turns to terror as he realizes that the dog is chasing him. And it's gaining fast.

Then the animal is on his rear tire. Snarling, it lunges for Hollan's right foot and snaps, but all it gets is air, its fangs barely missing the pedal. Hollan's heart is galloping as he glances back at the beast. This close, he can see it's not a dog. It looks to be closer to 150 pounds. Dogs aren't this big.

It's a Canadian gray wolf.

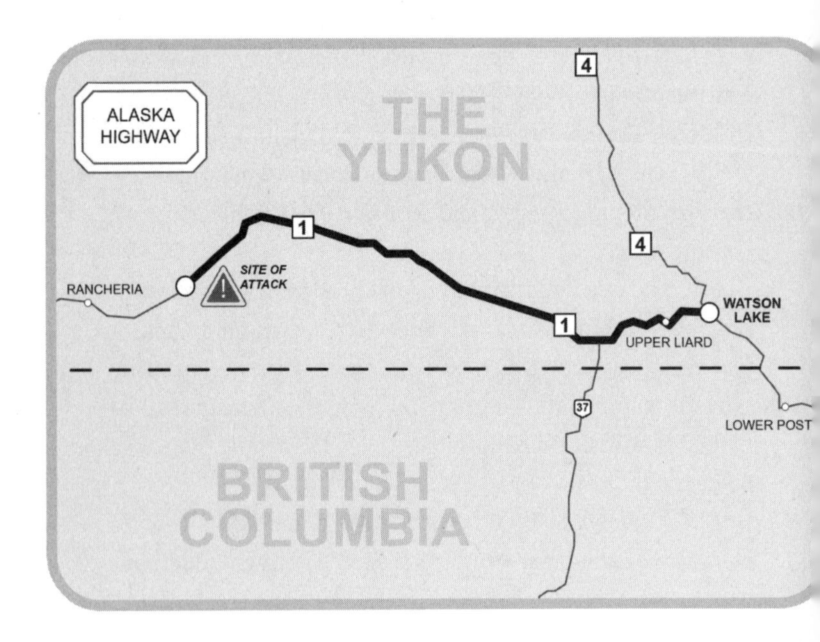

> When you panic from stress, the part of your brain that includes your amygdala—the fear center—becomes hyperactive. "You want the amygdala to respond and start the fight-or-flight process," says Olga Terechin, a medical doctor and psychiatrist. "The prefrontal cortex—which regulates your higher-level thoughts, actions, and emotions—then steps in and assesses the situation. That moment of cognitive assessment determines whether you'll be okay or you need to act now," explains Terechin. "Whether the threat is a wolf or a dog is being assessed by the prefrontal cortex. Once the brain realizes it's not a dog, the prefrontal cortex coordinates with the amygdala to act more instinctively."

And it's not giving up the hunt. Hackles up, ears forward, the wolf sprints at astonishing speed, easily matching Hollan's twenty-mile-per-hour pace. It bares its teeth as it runs, waiting for another bite at Hollan.

Feeling a surge of adrenaline, Hollan shifts his bike to the highest possible gear and stomps on the pedals. They're on relatively flat terrain, so he's able to hit racing speeds—25–30 mph.

> Getting chased by a lone wolf is rare. You're more likely to be stalked by a pack. In either instance, your body will respond with a rush of adrenaline and a heavy dose of panic. Adrenaline is a stress hormone, also called epinephrine. It comes from your adrenal glands, located on top of your kidneys.

> Adrenaline surges when you're under a perceived threat, and it's responsible for your fight-or-flight response. When released, an adrenaline rush happens in mere seconds, triggering several responses from your body.

But he's not pulling away from the wolf. He leans forward, poised over the handlebars, pumping his legs furiously. Then he remembers that he's carrying bear spray. He unzips his handlebar bag and pulls out the canister.

> During an adrenaline rush, your heart rate will increase (your heart will feel like it's racing); more blood is pumped into the muscles, giving you a burst of strength and energy; your digestion slows so that other parts of your body can get more blood; and you breathe faster, to maximize oxygen intake. Adrenaline relaxes your airways to increase oxygen too.

Still pedaling, he snaps off the safety with a free hand and points the nozzle toward the wolf. He glances past the animal, back down the road behind him. Still no sign of Gabe or Jordan. He mashes the trigger, issuing a cloud of irritating capsaicin—strong pepper spray that stings and confuses—directly at the beast's snout. Stunned, the wolf drops back twenty feet. But it quickly resumes the chase.

Hollan tries to use the seconds to put some distance between himself and the wolf. But the animal is closing the gap again. And then, like a horror movie scene on a hideous loop, the wolf is once more inches from him. Its jaws fly open as it leaps at the bike. They snap shut and this time they make contact—not with Hollan's leg but one of his pannier bags, the pouches on either side of his rear tire that contain his tent and camping gear.

The wolf tugs, forcing Hollan to shift his weight to keep the bike upright. Its fangs slice the bag open, and tent poles clatter across the highway. Hollan aims again with the bear spray. Pedaling madly, it's hard to get a bead on the wolf, but it's his only defense. Another quick blast, and the wolf recedes.

With this adrenaline coursing through you, you think more clearly and quickly. The hormone helps your brain sharpen mental focus. You'll be able to form a solid plan faster in this state. "The brain uses glucose to function, so adrenaline increases your blood glucose, giving the brain more access to energy for the brain cells," says Terechin. "Adrenaline also causes release of excitatory neurochemicals to speed up your brain function."

Turning a bend on the highway, Hollan sees a tractor trailer heading his way. He's saved. He waves his arms wildly and screams, pointing back at the wolf just twenty feet behind him. The truck driver lets off the gas; Hollan can hear the brakes clamp as the truck slows. If he can just scramble off the bike and into the cab fast enough . . .

Except once the truck driver gets a look at what's unfolding on the shoulder of the road, he doesn't stop. The driver floors it past Hollan, who can feel the wind from the rig as it speeds past him.

"No!" he screams. He looks back. The wolf is making another push, the distance between man and beast dwindling once again.

This section of the road starts to climb—not ideal for trying to outrace a wolf. Hollan downshifts and pedals faster. His leg muscles burn, but he can't slow down—the wolf's keeping pace. A car appears in the distance, headed toward Hollan. He shrieks and gestures over his shoulder at the wolf, now mere feet from his back tire. Hollan can see the stunned driver's face before the car speeds off, just as the truck had. Three more vehicles pass. None stop. Despair mounts as each car drives by.

A horrendous cat-and-mouse game is unfolding with the wolf. It races up, just to Hollan's rear, gets a snoutful of pepper spray, drops back for a short spell, and effortlessly returns. The can of bear spray is getting lighter, and the wolf doesn't seem anywhere close to tired. Hollan wonders how many bursts are left.

Rounding another bend, Hollan's heart sinks. The next half mile of road is a steep climb. There's no way he can stay ahead of the wolf now. His legs are burning—he's never worked them so hard. He decides his best option is to get as far up the hill as he can before his legs give out. Then he'll dismount and use his bike as a shield to jab the wolf away and hit it with the last of the bear spray, if there's enough left.

One last push now. *Is this how I'm going to die?* he thinks. He imagines the pain when the wolf's teeth pierce his flesh.

> Should the worst happen and you become injured, adrenaline will also decrease your perception of pain. The pain is still there, but adrenaline distracts you from that sensation, allowing you to focus on the threat at hand. "Our internal opioid hormones release during trauma to relieve the pain and shift the brain's attention to survival," says Terechin. "During life-threatening situations, living through the event is the most important element. Once the threat is gone, those natural painkillers recede and your attention is allocated to the pain source because that's now the priority. The pain is signaling the brain that something is wrong."

He shudders and pushes harder. He hears the wolf's panting grow closer with every pedal revolution. It's as if he can feel the animal's hot breath on his heels.

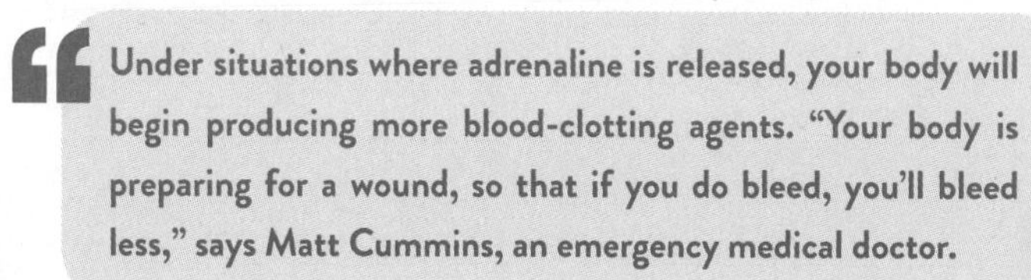

> Under situations where adrenaline is released, your body will begin producing more blood-clotting agents. "Your body is preparing for a wound, so that if you do bleed, you'll bleed less," says Matt Cummins, an emergency medical doctor.

He's a quarter mile from the start of the hill when an RV lumbers into view, coming down the hill, directly at him. *This is your last chance.* Moving the bike to the center of the road, Hollan screams until his throat burns. The RV approaches, slowing. Will it stop? Hollan's heartbeat thuds in his ears as the vehicle weaves by him. Then brakes screech as the RV jerks to a halt.

Unclipping his shoes from the pedals, Hollan vaults over the handlebars and sprints to the RV, never letting go of the bear spray. He hears his bike hit the pavement right as he reaches the back door. He yanks, hard. It's locked. Still screaming, he rushes to the passenger door, seeing the window is open. He starts to climb in through the window when the driver reaches over and opens the door for him.

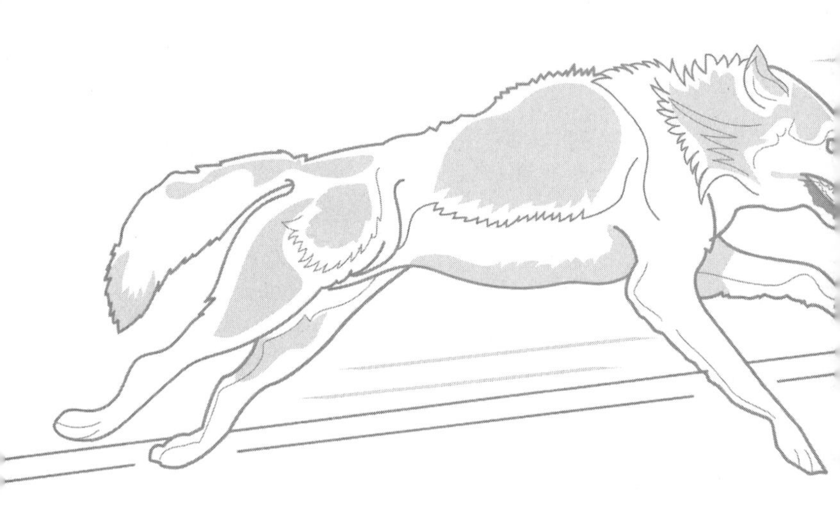

Leaping in, Hollan slams the door shut. The door's side mirror reveals that the wolf is already gnawing on his bike's shredded pannier bag.

* * *

Hollan and the RV's two passengers—a husband and wife—stare through a window, bewildered, as the wolf refuses to leave the bike alone. Other cars stop, blaring horns and shouting through cracked windows. Nothing spooks the wolf.

A woman jumps from one of the cars and flings a water bottle at the wolf. It's startled and retreats a few feet. Emboldened, a man exits another car and throws rocks. The projectiles force the wolf back, down into a ditch. A final throw lands squarely on the wolf's head. Confused and panting, the wolf's had enough. It slinks away, disappearing into a thicket of trees.

The couple in the RV let Hollan catch his breath. His friends arrive, Hollan's shattered tent poles in hand. After a crushing hug from the woman in the RV—one that feels like the greatest of Hollan's life—Hollan and his buddies resume riding. The RV turns around and follows for a few miles, watching the riders' backs. Just in case.

Ten miles down the road, the adrenaline subsides. Dizzy and tired, Hollan feels like he's moving in slow motion. Dismounting near a roadside creek, he staggers to the water and collapses in it. For the next fifteen minutes, Hollan sits in the water, shaking and swearing, trying to make sense of what just happened. And how close he came to fighting to the death with a wolf.

> Adrenaline rushes can last for up to an hour after an incident such as being chased by a wolf. The crash from the hormone subsiding is exactly what Hollan experienced: dizziness, lightheadedness, confusion, and blurred vision. You may feel jittery, a by-product of extra blood in your muscles, and some people become irritable. "Adrenaline surges cause abrupt changes in your body functions," says Terechin. They effectively disrupt what your body was doing before the incident. "When the threat is gone, everything slowly returns to normal," adds Terechin. "Your glucose and blood sugar goes down; the heart rate and your blood pressure decreases. It's like when you do a hard workout and your heart rate spikes, then you stop and feel a little dizzy. It's a major circulatory shift."

HOW TO SURVIVE WOLVES STALKING OR CHASING YOU

LOOK FOR THE PACK. Wolves mainly hunt in groups. If you see a wolf, turn slowly to see if others are nearby. They likely are, and they're probably circling you, trying to corral you.

KNOW WHEN A WOLF IS BEING AGGRESSIVE. It's unusual for wolves to be aggressive toward humans, but you should know the signs. Ears pointed up and forward mean it's trying to get the best sense of your location, while a lowered head helps it get a better sightline on you. Raised hackles—the hair on its back between the front shoulders—are another indicator that the wolf is aggressive.

DON'T RUN. If facing a wolf, do not run. Wolves can run for short bursts up to 37 mph. Instead, make solid eye contact and back away slowly, always keeping the wolf or the pack in front of you. Running can trigger the prey drive within a wolf and cause it to chase you when it otherwise would not have. This is perhaps why the wolf chasing Hollan wouldn't relent. At lesser speeds, wolves have been known to chase prey for hours, waiting until the target tires before moving in for the kill.

BE AGGRESSIVE AND LOUD. If the animal is approaching, clap, shout, throw stones or sticks; do anything you can to seem like a threat. If you have an air horn or noisemaker, use it. Let the animal know you are not easy prey. Grab a rock or something heavy, in case you need to defend yourself.

CARRY BEAR SPRAY. This deterrent helped save Hollan's life. Use short bursts, aimed at the wolf's face. Resist the urge to unleash the whole can—they can empty in as little as seven seconds. You will likely need several sprays to keep the wolf at bay.

IF ATTACKED, FIGHT BACK. Don't curl up into a fetal position or play dead. Wolves don't care, and it won't stop the attack. Instead, use anything as a weapon and strike the wolf with everything you have. Go for the eyes and snout. If you're able to, back up against a tree or something sturdy to protect your back and give you some support.

CLIMB A TREE OR ROCK. Wolves can't climb, so you'll be safe if you're able to get up high. Wait a long time before coming down; wolves are smart and have been known to lurk in the area, hoping for another chance to attack. If there's a fire nearby, run toward that. Wolves hate fire and smoke.

SHOVE YOUR FIST IN THE WOLF'S MOUTH. Yes, this sounds insane, but some biologists believe it prevents the wolf from biting you. It'll stun the wolf and make it harder for the wolf to breathe. If it can't get adequate breaths, it'll likely decide you're not worth continuing the attack.

SURVIVAL 101:

ATTACKED BY ANIMALS

Fighting another living creature is terrifying, particularly if that beast is double or triple your size, like a ferocious great white shark or enraged saltwater crocodile. But there are a few things to remember if you're facing down an apex predator in the wild.

1. ATTACK ANIMALS' VULNERABILITIES

Even in the most armored of animals, there's always a weak spot or vulnerability that can be exploited. As Leeanne Ericson and Craig Dickmann learned, it's almost always an animal's eye, so do whatever you can to inflict maximum damage there. Another decent option is the nose or respiratory system (stick your hand in a shark's gills and rip); if an animal can't breathe, it can't attack you.

2. ADRENALINE IS YOUR FRIEND

Responsible for fight or flight, adrenaline will get you moving, help you react faster, boost your strength, protect you from feeling pain, and allow you to get through the traumatic attack. Embrace any instinct that may arise in the moment. Your body—driven by adrenaline—will guide you on how to survive.

3. STEM MAJOR BLOOD LOSS

If you're going to be in the remote wilderness or somewhere that has the potential for a serious animal attack—like Bart Pieciul in Alaska's mountains—bring a tourniquet and/or quick-clotting powders. If you're bleeding heavily, you may not survive long enough to reach a hospital, so stopping major hemorrhages right after the attack is key.

4. PROTECT YOUR HEAD AND NECK

Bears, sharks, crocodiles, and even wolves can do serious, irreparable damage with a single bite or blow to your head or neck. There are important nerves and arteries in these areas, so whenever possible, shield these areas during the attack.

5. TIME IS TISSUE

When our skin, soft tissues, and organs are damaged, oxygen isn't flowing to them. The longer cells go without oxygen, the more they die. As Jeremy Sutcliffe learned, the body can begin shutting down rapidly after a rattlesnake bite, and Craig Dickmann had a limited window in which to save the skin on his degloved hand. If you're going off the grid, bring a satellite phone to ensure you can call for help instantly.

PART III

NATURAL DISASTERS

When Mother Nature unleashes devastating wildfires, torrential flash floods, and violent tornadoes, our fragility becomes crystal clear. In the blink of an eye, everything around us becomes an enormous threat to our survival. In the aftermath of an earthquake or avalanche, we're often on a short clock to be rescued. Acting quickly when a natural disaster strikes can be the difference between life and death.

BURIED BY AN EARTHQUAKE

11

JANUARY 26. 8:35 A.M.

His father's booming voice snaps Viral Dalal awake: "Get up!"

Dalal grumbles. "Please, Dad, I need a few more minutes." The twenty-four-year-old is stretched out on a mattress on the floor, the only space available in this crowded second-floor apartment in Bhuj, India, where he's on vacation with five members of his family.

His father nudges him again. "Fine, but move to the bed. I need to put your mattress in the other room." His dad points to a vacant bed that Dalal's brother occupied the night before. Dalal gets up with a groan and shuffles to the bed, wrapped in a comforter like a caterpillar. His father closes the blinds and tosses another blanket onto his son.

But as he starts to drift off again, Dalal hears a rumble. He feels it too. It's like thunder, but stranger, and growing louder. The floor begins shaking, and the bed trembles. The nearby dresser skitters across the floor.

The rumble quickly becomes a roar as the shaking intensifies. Suddenly, everything goes airborne—the dresser, a metal cupboard, the bed with Dalal still in it. It's as if they're free from gravity's pull—but in a split second, gravity returns to slam everything back upon the trembling floor. Again it happens. And again. Sounds of shattering glass rise over the rumble. The shaking is beyond violent now, like a series of explosions erupting from below.

Earthquake! Dalal throws off the comforter and tries to get out of bed to make it to the door. But each time he rises, he's flung in a different direction. Chunks of plaster and concrete start raining from the ceiling. Walls give way, falling inward in large chunks. A six-inch-wide crack opens in the ceiling, exposing steel rebar within the concrete. Dust fills the air. From the next room, Dalal's mother screams.

A loosened slab of the concrete ceiling, as big as a car, teeters overhead, threatening to crush him. He rolls onto his side, seconds before it crashes down. He's uninjured, but the whole ceiling is collapsing. His world goes black as he is tossed around, bumping into the concrete slab that nearly killed him. Fifteen seconds later, the shaking stops. Dalal can see nothing but dusty darkness.

During a big earthquake, a single blow to the head or neck from falling debris can cause death, due to skull fractures or brain bleeding; or it may snap your neck and asphyxiate you. There's little you can do to prevent this beyond covering your head and neck.

"Mom! Are you hurt?" he shouts.

The words barely leave his lips when he feels himself sliding. The entire room is on a tilt. Then he realizes it's not

just this room. The entire eight-story building is toppling. He reaches out to grab hold of something, but everything has come unmoored. He's falling now, lost in a cloud of cement chunks, broken bricks, and glass.

Dalal lands so hard on something that it knocks the wind out of him. He's flat on his back now, arms above his head in loose rubble. But still the building continues to implode around him. A concrete slab falls and stops just two inches above his face.

> "If, after the quake has subsided, you find yourself pinned under large debris, such as a wall or ceiling, any crush injuries you may have sustained can be life-threatening. When tissue and bone undergo prolonged compression, the crushed areas can die rapidly, requiring amputation. Releasing that pressure, however, can cause crush syndrome, which can kill you. Excess myoglobin—a protein in your skeletal and heart muscles—and other decaying cells from the injured muscle and tissue flood your kidneys, causing renal failure. If victims are under pain management, they may be happy right after rescue, then die minutes later. This is sometimes called a "smiling death.""

For a few seconds, it's eerily quiet. All Dalal hears are his own gasping breaths. Then, a series of deafening bangs, coming from overhead. Dalal realizes that the upper floors of the eight-floor building are pancaking down on top of him. He frees his hands and tries to press upward on the slab above him, afraid he's about to be crushed, but it won't budge. He's trapped. And if the slab above him jostles loose, he's dead.

But for now, anyway, there's stillness. It's over.

Miraculously, he doesn't seem to have any injuries. He takes a breath and screams out for his family, each by name. His father, mother, brother, sister-in-law, and two-year-old nephew. Dust fills his mouth every time he opens it, causing him to gag. Wiggling his hands to his chest, he's able to use the puny light on his wristwatch to see his tomb. Stifling concrete slabs surround him. He has four inches of space above his knees and feet, and jagged debris on either side means he can't push his arms or legs out. His head can rise two inches, but no more. The concrete prison is barely wider than two feet. His hands explore below him and feel a bedsheet. Somehow, he's still on the mattress.

Now what?

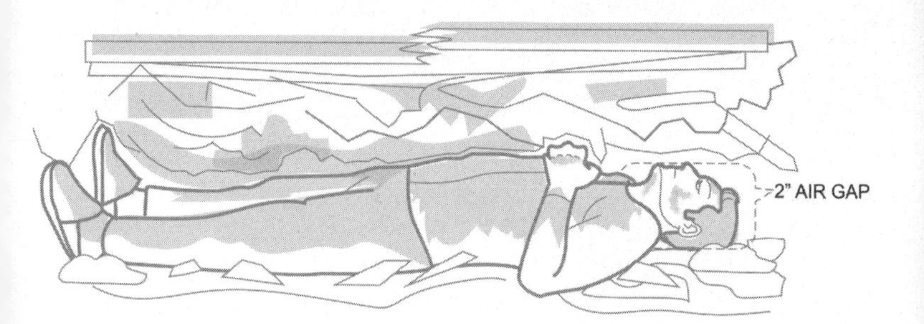

* * *

A 7.7 magnitude earthquake had just leveled Bhuj, a city of 150,000 in western India. Not only does Viral Dalal not know the extent of the destruction; he can't even accept that the apartment building in which he was sleeping just moments ago has been reduced to rubble. He doesn't think this is reality; he thinks it's a dream.

> Dalal's initial reaction of denial is completely normal. "Our brains operate in a predictive model," explains Alex N. Sabo, a medical doctor and psychiatrist. "It estimates what it thinks will happen next, so when that reality doesn't align with what we're seeing and experiencing, such as an earthquake, it'll lead to brain shock." Brain shock involves a massive release of noradrenaline that will take the frontal lobe (the part of the brain responsible for nuanced thinking) offline, and it will activate primitive survival mechanisms that are unconscious at first to freeze, fight, or flee. "The shock experience is remembered in the body as a shudder or shiver, a wave of heat or cold or a hollowing out of the insides," says Sabo. "And it will be triggered in the future by the sounds or touch, the trembling of the earth, for instance, that occurred as the earthquake began."

"How can this happen?" he screams. The words bounce off the concrete and echo in his ears. What about his family? Are they also trapped? Are they even alive? He tries willing himself to sleep, thinking that when he wakes up his nightmare will be over. Then a numbness washes over him as shock sets in. *I'm buried alive. And there's no way out.*

He works his hands under his back. Bits of brick and glass shards from a broken overhead light had gotten between him and the mattress during the collapse, and they're poking his back. It takes time, but finally he clears them away.

The air already feels stale, and Dalal wonders how many cubic feet of it he needs to breathe every hour in order not to suffocate. When he inhales deeply, his expanding chest hits the slab above him. He decides it's best to take shallow breaths, unsure if his air will soon vanish.

Negative thoughts cascade in his brain. *Will an aftershock loosen the concrete slab above me and crush me to death?* He forces despairing notions out; those are the enemy of survival, he says aloud. He'll focus only on positive, helpful things.

> **When trapped in a confined, dark space, the brain will start exploring itself, according to Sabo. "Your mind will think of death and pain and anger. It'll be overwhelming," he says. "And negative thoughts will trigger more negative thoughts."**

Except for the wristwatch's tiny light, which he uses sparingly, Dalal lies in complete darkness. It's overwhelming and he's feeling panicked, so to distract himself, he uses the watch to locate a small piece of metal—a bit of twisted light fixture. He turns off the light and starts to scratch at the concrete beside him, a bid to find the phone that was near the bed. After an hour, he turns on the light to examine his progress. Frustratingly, it did nothing. Then the ground rumbles again. It's the aftershock he feared. His body shakes violently as he prepares for the slab to plummet down and crush him. But the slab doesn't budge and the aftershock passes.

Hours pass. No sounds. No anything. He tracks the passage of time on his watch. At the fourteen-hour mark, he hears his stomach growl. It's coming up on twenty-four hours since he last ate or drank. The pitch-black surroundings are making him mentally uncomfortable. He wants to stretch, to stand, to do anything. And he can't. The only thing he can control—sort of—is his brain.

Shock exists for a number of reasons, but the main one is to ensure we live past the event. "Shock reduces our physical and emotional pain and helps us dissociate from the event," says Olga Terechin, a medical doctor and psychiatrist. "How we dissociate isn't known with certainty. Overall, our perception and the body's proprioception—a series of signals that feed the brain information about our body's movement, location, and position—is not as reliable as we think," explains Terechin. "Under stressful situations, as Viral experienced, you can feel like you don't belong in your body or that your surroundings are not real."

And even that doesn't always cooperate. He works to keep thinking of fond memories, such as favorite meals, places he's visited, and laughing with his family, to avoid intense feelings of panic and anger. Somehow, he falls asleep.

> We can think away pain too. "There's cognitive relearning therapy that's effective for people with chronic pain," says Beth Palmisano, a medical doctor who specializes in pain management. "The aim is to shift attention away from the pain, erratic thoughts, and fears to more adaptive thought patterns." It can work in situations like Dalal experienced too. Thoughts and beliefs will influence physiological reactions, Palmisano says, noting that negative thinking can increase pain perception. "People who associate pain with something larger than themselves can withstand more suffering," she says. "If Viral thinks he needs to keep going for his family, that focus can help shift his perception of pain."

* * *

He feels weird waking up in absolute blackness, no sounds of birds or bustling Bhuj traffic. It's the second day in the rubble, very early in the morning. He hasn't run out of air yet so he breathes normally. He's starving, so he imagines his favorite foods to make his mouth water, swallowing whatever saliva he's able to produce. It quells the hunger pangs, but only momentarily.

> What Dalal experienced when trapped was sensory deprivation. Losing our senses as well as the ability to control our environment is highly detrimental to our mental state. "The brain is locked in our skull without any communication to the outside world," says Terechin. "It predicts and constructs about what's going on outside, but it uses information from our senses to do so. When we can't hear or see, we're not giving the brain any input to make an educated guess." But the brain still tries to come up with an explanation for the environment. "It'll make up noises or visuals, because the brain believes there must be something happening that we're not sensing."

Dalal tries a new tactic. He hates eggplant, so he imagines that eggplant is the only available food. The trick works; the more he thinks of what he abhors, the more okay he is with being hungry.

Aversion thinking doesn't work as well with thirst. He tries thinking about drowning in a flood. Then he thinks about the movie *Titanic*, imagining drowning after abandoning ship. It doesn't help.

It's been forty-two hours since he used the bathroom, and he doesn't want to waste the only liquid to which he has access. Dalal's fingers find a small dome-shaped electrical covering that's attached to one of the slabs. He pries it off and uses it to collect—then drink—three small portions of his own urine. It tastes vile, but it helps.

> How do you mentally prepare to drink your own urine? "It's rationalizing," says Terechin. "Ever look at an oyster before you eat it? You can talk yourself out of being disgusted by anything, even drinking urine. Our aversion to our own waste—feces or urine—is to keep us away from it to prevent getting sick by bacteria. We're able to overcome this aversion with a certain amount of internal dialog, believing that the benefits will outweigh the risks. And then you go for it."

* * *

The first thought Dalal has on the fourth day is a dark one: *Why even bother waking up*? Exhausted and sore, it's harder to suppress negative thoughts. He heard some pinging noises yesterday, ones that sounded man-made. But they stopped quickly. He's heard little else. He's still drinking his own urine, but the bitterness in his recycled waste has grown tenfold. He chokes down a meager portion now and wonders how much longer he can survive like this.

> Dalal's decision to drink his own urine is ill advised. "The body excretes waste, putting salt and toxins in the urine," says Terechin. "Urine is trying to get rid of salt. When you drink it, you're consuming the salt again, and that will further dehydrate you. You lose moisture as you breathe, so you're only increasing your dehydration. Drinking your own urine is not recommended."

Dalal's back is so stiff, he's not sure he'd be able to get up even if this slab is removed. The big toe on his left foot is starting to go numb. All his muscles ache. His neck hurts the worst; he can only lift his head a few times an hour now without immense pain. He stretches each limb, then his back. It feels a little better. In the evening, he hears dogs barking. Dalal screams until his throat is raw and his voice raspy. But the barking fades after a few minutes.

> Dalal's body ached from being immobile for days. "Pain thresholds are complicated, modulated by factors like genetics, gender, culture, and even training," says Palmisano. "When you train and prepare for stressful situations, such as combat, your body can acclimate and the pain perception can be less, because it's expected." For regular people under heavy stress, such as being buried alive in an earthquake, pain perception is usually increased. "However, when we're subjected to a constant painful stimulus, the brain can respond by gradually reducing the body's response," says Palmisano. "It's a defense system that makes things more tolerable."

Back into the silent darkness.

* * *

Around seven in the morning on the fifth day, Dalal hears something. Human voices. People shouting; it's muffled but distinct. Then engines revving, machinery beeping. Next, a big rumble. Not an aftershock—it sounds more like debris being cleared away, just beyond his feet.

"Help!" he shouts. "Anybody there?" Dalal repeats this cry every few seconds.

"Quiet!" A voice from outside. "Who is there?"

Dalal shouts his name and hears it being repeated outside. A calm washes over him. *I am going to be free soon.* But then a new worry: What if mistakes during rescue efforts lead to the slab smashing him? He yells this to the crew outside and suggests they start near his feet. When he hears tapping in that vicinity, he confirms that's the right spot.

Sounds of concrete being chipped away fill Dalal's confined space. Then he hears metal rods snapping and debris being cleared. After half an hour, there's a sliver of light.

"I see daylight!" Dalal yelps.

Slowly, the hole widens. Rescuers carefully place supports inside the hole, to prevent the slab from falling on Dalal. More minutes pass. Then, a sensation on his foot: Someone's touching it. Down near his knee, Dalal sees it: the hand of one of the rescuers. He clasps it tightly.

"We've got you, Viral," a voice from the outside says. "You're going to be okay."

* * *

The 2001 Bhuj earthquake lasted ninety seconds and killed more than 20,000 people, injured another 167,000, and demolished 340,000 buildings across eight cities in India.

Tremors were felt as far away as Bangalore, more than 1,100 miles to the south.

Viral Dalal spent one hundred hours buried under the debris from the collapsed apartment building, under a pile of rubble forty feet high. When he was pulled from the rubble, feetfirst, he didn't have a single scratch.

Dalal was reluctant to go to the hospital to get checked out; he wanted to stay and help locate his parents, brother, sister-in-law, and their two-year-old. Forced into an ambulance, Dalal returned to the rubble an hour later. He remained there until the bodies of his entire family were recovered. Of the six members of the Dalal family buried in the rubble, only he survived.

HOW TO SURVIVE AN EARTHQUAKE

If you're outside when an earthquake starts, stay there. Move away from buildings, utility lines, and any other structures. Lie down and wait out the shaking. If you're inside during an earthquake, there are several things to know that could save your life.

DROP. COVER. HOLD. Get on your hands and knees before the earthquake can knock you down. Then get under a sturdy table or something that can protect you from falling debris. Nothing around? Cover your head and neck with your arms. Hold on to the protective furniture—or your head and neck—until the shaking stops.

AVOID DOORWAYS. You're safest under a table. What's most likely to injure you is flying or falling debris—glass, cement, wood, or bricks—and doorways won't protect you from this.

MOVE AWAY FROM POTENTIAL DEBRIS. You'll have a few seconds after the shaking starts to move before it intensifies, and you may become stuck in place as Dalal was. Avoid being near anything made of glass or any furniture or items on walls that could land on you. Avoid cowering underneath anything on the ceiling too, like light fixtures or fans.

AVOID THE CENTER OF THE ROOM. If you can't get under a table, move toward bare walls. The center of the room is typically the most dangerous place to shelter during an earthquake, because if the ceiling collapses, that's the most likely spot for it to fall.

RUN IF THE BUILDING COLLAPSES. You'll have little warning, but there will be signs. As Dalal saw, the ceiling will crack and the walls will separate. When this happens, do whatever you can to get out. Run toward the light, if you're able. If the building is under three stories, windows are the best point of egress. Try to get to a stairwell in a taller building; they're sturdier.

PROTECT YOUR HEAD AND FACE. As the building collapses, cover your face and head with your hands, creating a buffer zone. As the debris settles, it'll give you an air pocket and some space around your head.

COVER YOUR MOUTH AND NOSE. Plenty of the debris and dust kicked up into the air will be toxic to breathe. If you're able, cover your mouth and nose with something made of cloth, such as a T-shirt.

STAY IN THE PRESENT. If you're trapped like Dalal, maintaining presence of mind is a challenge. "Being encased in concrete and darkness with inches to move would mentally challenge even the most well-trained Navy SEAL," says Sabo. Pair your thought process with the brain stem–driven process of breathing to help you stay in the moment, Sabo recommends. "Breathe in and say, 'I'm breathing in,' then breathe out and say, 'I'm breathing out.' Three rounds of this can focus the mind in the present, preventing you from thinking about the past or the future."

THINK POSITIVE. "Negative thinking leads to more negative bias about your outlook," says Terechin. "It will be difficult to focus on the fact that the sun is shining, because negative thoughts feel congruent with your state of emotion and physical predicament," she says. "Try to focus on positive memories, as Viral did, to keep from spiraling."

THINK AWAY HUNGER. Our consciousness is capable of believing it isn't hungry by thinking of hated foods. "You need to believe it, though," says Terechin. "Dive into the thought and repeat it to yourself. Averse memory coupling activates an area of the brain responsible for disgust, so when you connect hunger with emotional disgust, it can tamp down your hunger feelings—just at that moment." But Terechin notes that Dalal's feeling of hunger would come back as soon as he stopped thinking about eggplant.

ENGULFED BY AN AVALANCHE

FEBRUARY 19. 11 A.M.

Barreling through knee-deep powder, Elyse Saugstad can't help but smile. Backcountry skiing is where the world-champion freeskier feels the most alive. Forging her own path through virgin snow, she feels a deep connection to this remote, rugged terrain. The snow sprays around the thirty-three-year-old as she carves down Tunnel Creek, a slope on the southwestern side of Cowboy Mountain in Washington State.

It had been a slog to get to this pristine area, requiring two chairlifts, then hiking past a boundary gate and several ominous signs, including: SKI AREA BOUNDARY. NO SKI PATROL OR SNOW CONTROL BEYOND THIS POINT. CONTINUE AT YOUR OWN RISK. At the gate, Saugstad and her fellow skiers passed an electronic checkpoint, which made their avalanche beacons chirp, confirming they were active. Avalanche beacons are essential in the backcountry. They constantly transmit your location and, in emergencies, can be switched to receive the location of other beacons nearby. If someone gets buried in an avalanche, the beacons can be the difference between life and death.

In addition to her beacon, Saugstad carries another piece of lifesaving technology: an avalanche airbag. Deployed by tugging on a cord, this system inflates twin airbags on her back. It can protect her from trauma by serving as a buffer against hazards like rocks and trees, and it can help buoy her and keep her near the top of the snow. Between her survival gear and her experience, Saugstad feels safe as she slaloms through the deep, undisturbed powder of Tunnel Creek—even though the avalanche forecast for this February morning registered as medium, verging on high.

Saugstad comes to a halt after a sprint down the steep hill. Elsewhere on the slope, others in her group are picking their lines, making good time down the top part of the mountain, called the headwall. Another skier comes to a stop beside her: Chris Rudolph, marketing director for Stevens Pass, the ski resort on the other side of this mountain. Rudolph is her friend and guide—and the other reason she's not worried. He knows Tunnel Creek intimately, having made hundreds of runs here. He'll get her safely to the bottom.

Rudolph beams at Saugstad. "Isn't this incredible?" he says. "Tunnel Creek is the best." The area is renowned to locals as being a powder stash—an unregulated freeskiing playground the likes of which this expert crew dreams about. And conditions today are perfect: During the past two days, storms have dropped twenty-six inches of fresh snow, but today it's partly sunny and calm. "It's amazing," she replies.

She watches as two other skiers, Rob Castillo and Johnny Brenan, come flying down the slope, skidding to a stop about twenty-five yards below them. One of them lets out a whoop and shouts, "That was sick!"

Saugstad is about to ski down toward them when she hears Rudolph scream: "Avalanche! Elyse! Avalanche!"

Saugstad turns and looks up, back toward the summit, to see snow barreling through trees, a wall of white roughly two stories tall, advancing upon her at the speed of freeway traffic. *Move. Now!* her brain shouts. She tries to go right, to the side of the impending onslaught, but it's too late. The snow slams into her, knocking her off her feet. It's like being hit by a freight train.

Frantically, she yanks the cord on her avalanche airbag. In the chaos, as she's tossed around, she's not sure if it even inflates.

Everything in her field of vision goes white as the avalanche shoots her down the hill. She's being disassembled—the mass of snow yanks her ski poles from her wrists and tears her goggles from her face. Her nose ring rips out as snow fills her mouth and nostrils.

Her fear builds as the assault continues. It's been seconds, but it feels like an eternity. *This is it*, she thinks, flipping over and over. *This is how I die.*

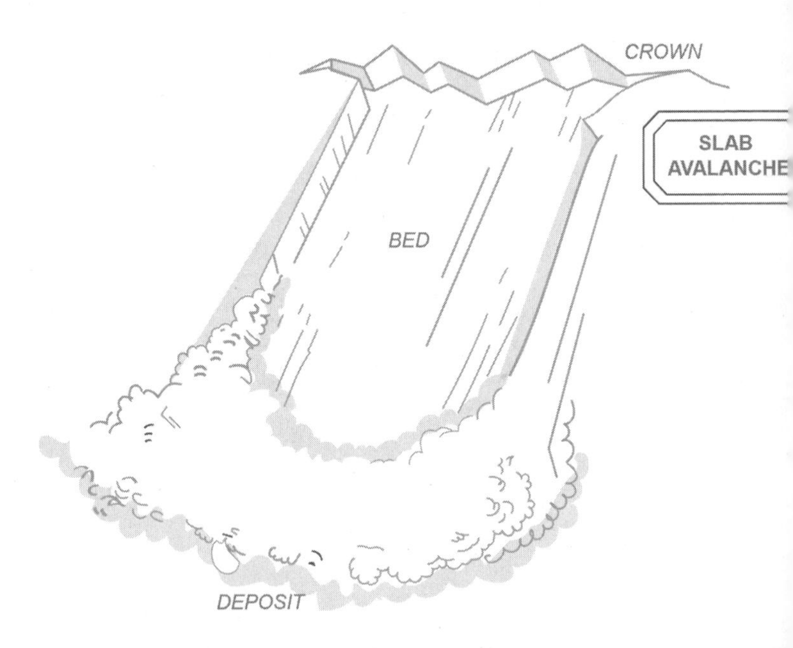

> **"About 10 percent of avalanche victims succumb to a combination of suffocation and head trauma, while another 5 percent die solely due to trauma. If you're tumbling around in the snow's wash at speeds beyond 20 mph, a single hit to a tree or rock can be fatal."**

* * *

Saugstad is enduring what's called a slab avalanche. They're the most common form of avalanche—and the most deadly. They're a common occurrence in Tunnel Creek due to the slopes of the mountain—typically 40 to 45 degrees. Such slopes retain massive amounts of snow, and they are steep enough for that snow to slide enormous distances if shaken loose. When heavy amounts of snow accumulate, and temperatures and wind rise, large swaths of that snow can break off in enormous slabs, many larger than the average city block.

Conditions are somewhat optimal for an avalanche this February morning, though murky enough to warrant an official avalanche forecast of "considerable." That's a tough designation for professional skiers, who know it's a gray area, somewhere between "proceed with caution" and "stay away." At daybreak, the crew was optimistic it would be the former.

It proves to be the latter. The slab that breaks off the top of Tunnel Creek is relatively small by avalanche standards, around 36,000 square feet, roughly half the size of a football field. But as it descends, the avalanche doubles in mass, attaining a final weight of close to 11 million pounds of snow, about the weight of one hundred tractor trailers.

Fortunately, when the avalanche strikes, most of Saugstad's group are above it, and others, anticipating the potential risk, have already skied out of its path. But those less fortunate will now have to fight to survive.

* * *

Rob Castillo is stopped, leaning against a spruce tree along the side of the trail, when he hears Chris Rudolph's scream. He barely has time to glance up and clock the tsunami of snow about to overtake him. The forty-year-old Castillo is a former pro skier, but no one can hope to outrace an avalanche like this one. His safest bet is to hook each arm around two small trees in front of him and hang on. He buries his head in between the spruce trunks and braces. He'd heard a weird hollow sound beneath the snowpack minutes before, and now he wonders if the mountain was trying to warn him.

A GoPro strapped to his helmet captures what happens next. The avalanche pounds him for sixteen seconds, pushing him forward. Castillo's shoulders are crushed against the tree trunks. His face smashes into the prickly pine needles. He groans and tightens his grip. If he lets go, he'll die.

The trees are now bending. Over the roar of the cascading snow, he hears other trees snapping. *If these break, what will I do?* Castillo tries to think of a plan if he's taken by the avalanche. But then, just when he's sure he'll be swept away, it starts to ease.

He can see daylight again, though snow continues streaming between his knees for another thirteen seconds. He blinks, calibrating his eyes, and peers up the hill, scanning for Rudolph, Johnny Brenan, and Saugstad.

No one's there.

He screams their names. No response. Just the eerie creaking of broken and swaying trees.

* * *

Inside the avalanche, Saugstad struggles to calm her racing mind. She bounces off something. It's a hard hit, maybe a tree or boulder. But it doesn't slow her descent in the slightest.

If she's going to stay alive, she must not panic. A surfing lesson she once had pops into her head: When the ocean drags you under, relax and save your energy until it eases. It *will* ease. She'll need every ounce of strength to stay alive when the avalanche stops.

She goes limp, like a ragdoll in a washing machine. Finally, she feels the torrent slowing, but she knows she's not out of danger. When the snow stops moving, it will freeze solid in a matter of seconds. She knows that suffocating in the snow is how most avalanche victims die. Her life depends on what she's able to do next.

> **Nearly 90 percent of avalanche deaths are due to asphyxiation. If you're trapped under the snow, the carbon dioxide you're expelling with each breath has nowhere to go, and you're breathing it back in. That toxic gas replaces the oxygen in your bloodstream, depriving your heart, brain, and vital organs, and your body shuts down.**

As the snow slows, she's disoriented and unsure which way is up. Still, she punches her hands outward toward what she hopes is the surface. As the avalanche stops, she tries to wiggle her limbs. Nothing moves. The snow is like cement and she's frozen in place. She can't turn her head. Her legs are locked beneath her, and the snow is so tight against her chest that she can't take in a full breath.

But her face is poking through the icy crust. Her airbag worked; it kept her more buoyant, so she wasn't completely buried. *Small victory there*, she thinks.

Her fingers are near her head and the surface. It's a struggle, but she frees her hands enough to pick at the snow entombing her face. As she claws away, her mind is flooded with negative thoughts: *Was everyone swept away? Is there anyone left to rescue me?* She's able to clear snow from her eyes and realizes her head points downhill. She takes in the view—an upside-down valley and mountain. *Please let there be someone.*

Her next thought sends a shiver through her immobile body: *What if there's another avalanche? I'll be buried alive.*

* * *

Two thousand feet up the slope from Saugstad, Wenzel Peikert frantically surveys the damage. Shards of thick trees, snapped like matchsticks, litter the ground. Everywhere he looks, the once-pristine powder is sullied, scattered with upheaved rocks, soil, and broken tree branches.

The twenty-nine-year-old part-time ski instructor was cutting across the slope, about to stop and rest, when he felt the avalanche starting. There wasn't a loud noise, just extreme wind and pressure and an immense rush of air. Miraculously, Peikert had crossed directly in front of the torrent. It missed swallowing him by inches.

Shaken, Peikert snaps into high alert. Time is precious for buried survivors. He knows many victims die of suffocation, which can happen in minutes. But he has to be careful—loose snow along the avalanche path known as "hang fire" can easily create a second avalanche without warning. After casting a wary eye up the mountain, he grabs his avalanche beacon and switches it to search mode, which allows him to find other beacons that might be buried. He starts traversing the slope, praying for a ping. But he hears nothing.

At high enough concentrations, carbon dioxide causes unconsciousness, followed by respiratory arrest within one minute. "Depending on how little oxygen you have, you can pass out in one minute, and die within five to seven minutes," says Deepak Sachdeva, an emergency physician. "In some severe avalanches, the snow can get into your airway, blocking it, and

that will kill you. If you're able to breathe in and out, then you have minutes." When respiratory arrest starts, it can be painful. "When your body recognizes that it's not getting oxygen, and knows there's too much carbon dioxide, it causes sensations like air hunger, and that's uncomfortable," says Sachdeva.

He moves downhill, toward a clearing in the dispersing fog. There, another member of the group is fumbling with his beacon. It's Rob Castillo.

"Are you okay? You were right in the middle of that," Peikert says.

"Yeah, I'm okay. I grabbed on to some trees and that saved me," Castillo replies. "But Elyse, Chris, and Johnny are gone—I think they're somewhere down the mountain."

The two men realize they'll need to work together to save anyone they can. They know at least three members are missing, but more could be in danger.

"Let's crisscross the avalanche path and look for a signal," Peikert says.

They begin skiing down the mountain slowly. After several minutes of fruitless searching, Peikert shouts to Castillo: "Listen, I think they're either up here, trapped in the banks, or they've been carried to the bottom. I'll go down there. If you don't find anyone here, come help me." Castillo nods and Peikert heads downhill.

It's more than two thousand feet to the valley floor at the bottom of a narrow, rocky gully.

It's unnerving in the wake of the avalanche; there's devastation wherever he looks. Twenty-foot walls of rock line either side of the gully, making him feel even more vulnerable. If another avalanche comes this way, he'll have no way to escape it.

His beacon is still silent as he pops out of the bottom of the gully. Here the scope of the disaster really hits him. The debris covers an area bigger than a baseball stadium, peppered with ice chunks as big as cars, rocks, and broken trees. It's too difficult to ski through, so he steps out of his skis and walks.

Suddenly, his beacon chirps.

He pans left and right. There's more than one signal. This is where all the victims are.

Peikert probes the snow with his ski pole, walking along as the beeping grows stronger. He looks up to see a tiny patch of pink atop the snow. *Is that . . . a glove?* He rushes toward it. *It is!*

"Help!" comes a voice near the glove.

Peikert drops to his knees and is surprised to come face-to-face with Saugstad.

"You okay?" Peikert says.

"I think so. But I can't move," Saugstad replies.

Peikert whips a shovel from his backpack and chops at the snow. It's excruciatingly slow; the snow is rock-hard. But others from the group soon arrive to help, and after several agonizing minutes, they finally succeed in freeing Saugstad. Peikert tugs her to her feet. She's scraped and bruised, but otherwise uninjured.

Dazed and in shock, Saugstad ambles around the debris field, her deflated airbag limply blowing behind her. But she's alive. Peikert knows that time is running out for other victims who may be buried. He leaves her and rushes off to help other members of the group, digging frantically at the sites of other beacons.

> **Data shows that 93 percent of avalanche victims survive if they're found and freed from the snow within fifteen minutes. After that, your chances drastically drop. Only 30 percent of victims are found alive after forty-five minutes; and after two hours, less than 5 percent survive.**

* * *

Rob Castillo is in the middle of the debris field when he hears Peikert shout that he's on top of something. Peikert gently probes the snow with a ski pole. After a few jabs, he feels someone. Both men race to dig. Soon enough a man's upper back emerges. The men quickly begin to dig where the man's head should be.

Except it's not there.

His neck is bent, his head tucked awkwardly underneath his arm. There's blood. The man isn't breathing. Castillo attacks the snow as fast as he can. Uncovering the body, Castillo realizes it's Johnny Brenan. Castillo begins CPR, shouting at Brenan's motionless body, trying to will it back to life as Peikert looks on.

For survivors dug out of avalanches, the first thing to do is check for a pulse. "The victim's heart rate can get very slow, so you may not be able to detect it easily," says Sachdeva. If there's no pulse, start chest compressions. If there is a pulse, then check for signs of respiratory distress. "You have to look, listen, and feel if they are breathing," says Sachdeva. "Do you see the chest moving? If not, get your face next to theirs—put your cheek an inch from their nostrils and mouth—and determine if you can hear or feel them breathing." If there's no breath detected, start rescue breathing. "Check the airway first, making sure it's free of snow or debris, then start mouth-to-mouth," says Sachdeva. "Pinch the victim's nose and cover their whole mouth with your whole mouth and breathe." Watch the chest wall to see if it's moving and expanding.

Saugstad watches Castillo beat on his friend's chest, but she knows the grim truth. Brenan didn't make it. She looks up the hill, where other skiers have found Chris Rudolph and are digging him out. Someone from that rescue effort shouts that Rudolph isn't breathing, and she can't believe her good friend and guide is gone. A few yards in another direction, another body is recovered.

Less than an hour ago, the group had been at the summit, all grins and laughs. Then the mountain unleashed terror, taking four skiers with it. Three perished. As she tries to piece it all together, amid the debris and heartache, Saugstad knows it's only through her preparedness and sheer luck that she is the sole survivor.

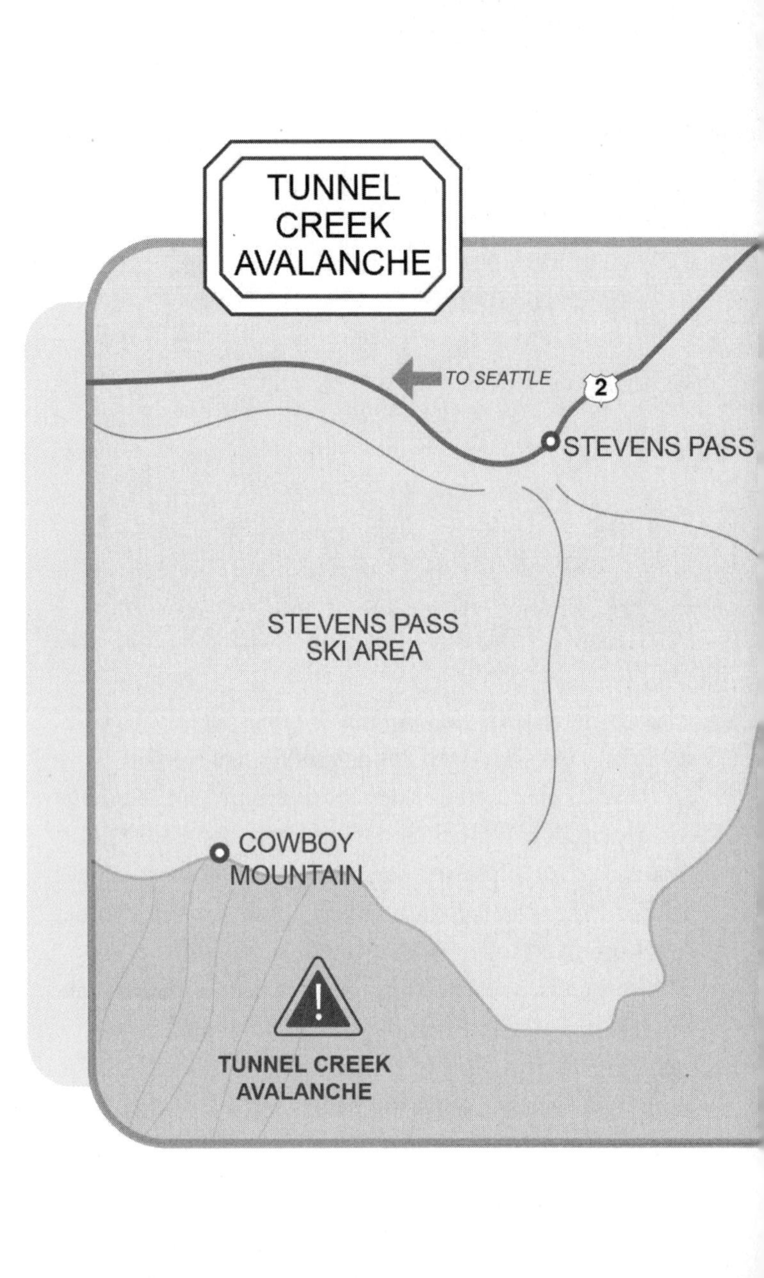
TUNNEL
CREEK
AVALANCHE
TO SEATTLE
2
STEVENS PASS
STEVENS PASS
SKI AREA
COWBOY
MOUNTAIN
TUNNEL CREEK
AVALANCHE

"Depending on how long you're stuck in the snow, hypothermia can be a concern. Any time your body dips below ninety-five degrees, it's considered an emergency, says Sachdeva. Some hypothermia signs include exhaustion, severe confusion, fumbling hands, memory loss, slurred speech, and drowsiness.

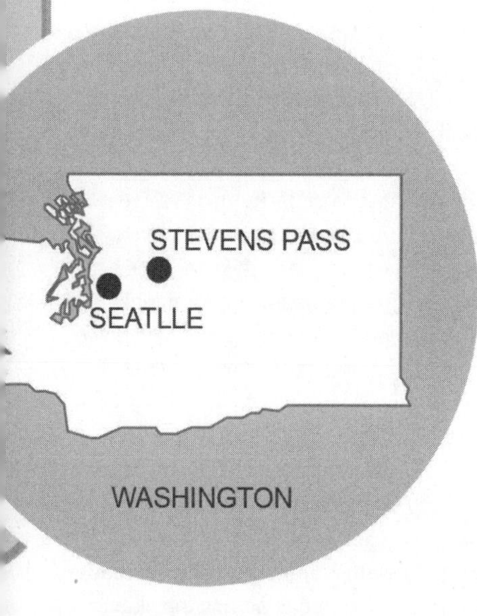

HOW TO SURVIVE AN AVALANCHE

BE PREPARED. Check avalanche reports often. Knowing when there's danger can help you avoid those areas. Avalanche risks can change hourly. Websites such as Avalanche .org can assist in getting the most up-to-date information.

DON'T GO ALONE. And make sure everyone in your group is packing lifesaving gear. If you're trapped in snow, you only have about twenty minutes to survive. Your rescue party consists of you and whomever you are with. There are three vital pieces of gear you need: a metal shovel (plastic won't work on hardened snow), an avalanche beacon, and a collapsible probe pole. Avalanche airbags, worn on your back, can help keep you floating near the surface. This works by increasing your overall volume; snow particles are smaller, so they sink beneath you.

LOOK AND LISTEN TO THE SNOW. Major warning signs include large fresh cracks in the snow, a high amount of windblown snow, and any weird or eerie noises coming from the snow. Most avalanches that prove fatal are slab avalanches, which is when an enormous slab of snow breaks apart like a pane of glass. When that happens, you'll hear a very loud *whumpf* sound, experts say. You'll have maybe a second or two to react.

SKI RESPONSIBLY. When on the mountain, be mindful not to cross above your partner, as this puts them at higher risk. Ninety percent of avalanches are triggered by humans. You can further improve your survival odds by identifying escape routes, searching for things to grab on to, and advancing down the mountain one at a time.

DON'T TRY TO OUTRUN AN AVALANCHE. You won't be able to; an avalanche can reach speeds of 70 mph or more within a few seconds. Instead, try to move across the slab to the side of the mountain, in what's called a slope cut. If you can't escape, then try to grab on to something: a tree, large rock, or whatever is nearby.

SWIM HARD. Can't grab on to anything? You'll have to ride it out. You'll tumble around like you're caught in a wave's undertow, but stay calm. As the snow slide slows, you want to swim as hard as you can to stay atop the snow. Your body is three times denser than the snow, so it'll start sinking. Kick and claw and do whatever you can to stay near the surface. Try to slide downhill feetfirst to protect your head and upper body from impact trauma.

DITCH YOUR GEAR. Make sure your skis or anything strapped to your body can be easily jettisoned, so they don't drag you down. Keep your backpack on, though, as that can protect your spine. Releasing gear as you slide can also serve as a visual trail that will make you easier to find.

ACT FAST. If you're buried, you have mere seconds before the snow hardens around you like concrete. This happens because as the snow gets packed together, the surface contact increases, and when points of ice touch, they fuse together. When you feel the snow slowing down, act quickly and do the following: Move a hand in front of your face to clear an air space to breathe easier; puff out your chest just before the snow stops moving, which will give you more room to breathe; and stick one hand straight up toward the surface. You can spit to find out which way is up. Many people are found in this manner.

MARK VICTIMS. If you see someone get caught in an avalanche, mark where the person is heading. Activate your avalanche beacon to receive and move in a zigzag pattern toward the victim; this increases your chances of overlapping with their beacon's signal. If the victim has no beacon, you need to probe with a pole. Look for higher mounds in the snow and probe there. Leave to seek help only after thirty minutes of being unable to find a victim.

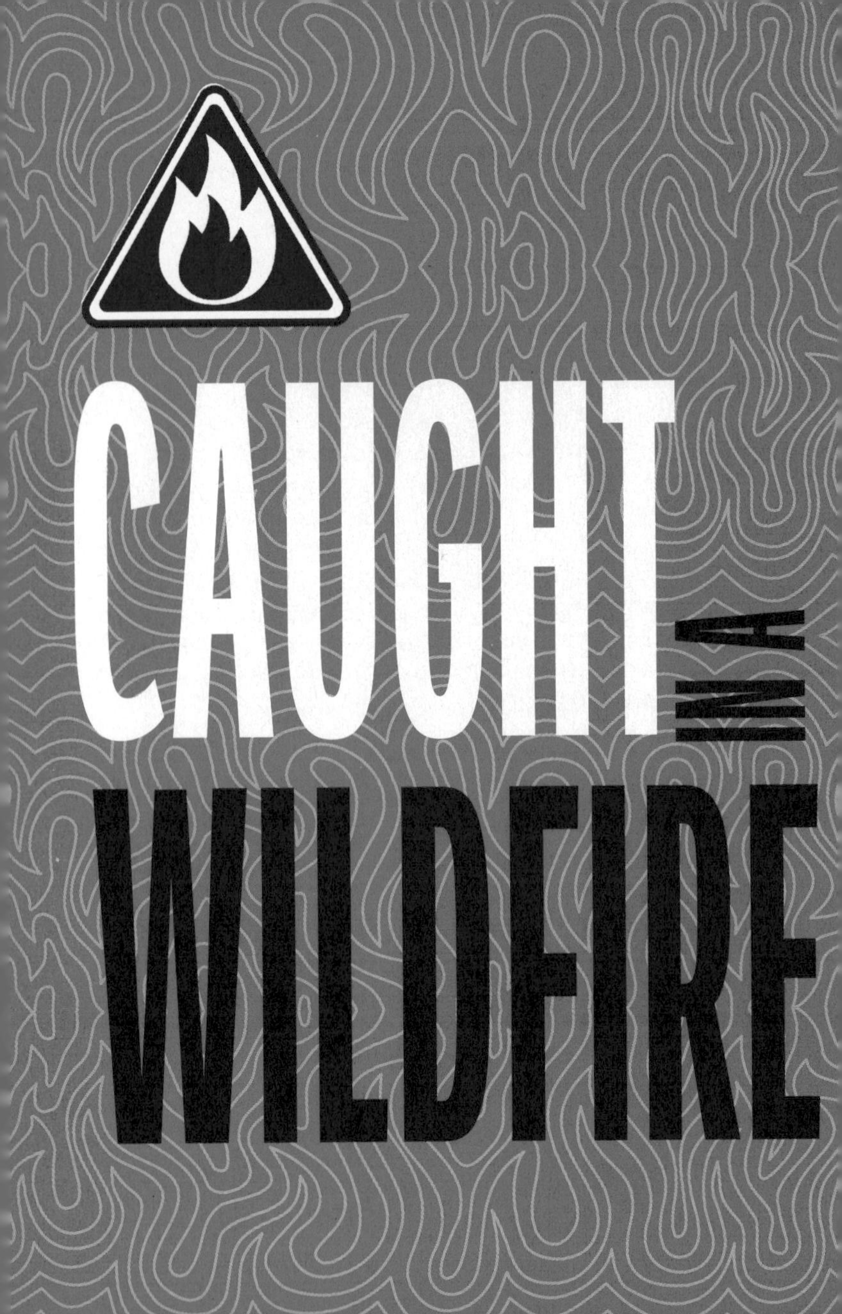
CAUGHT
IN A
WILDFIRE

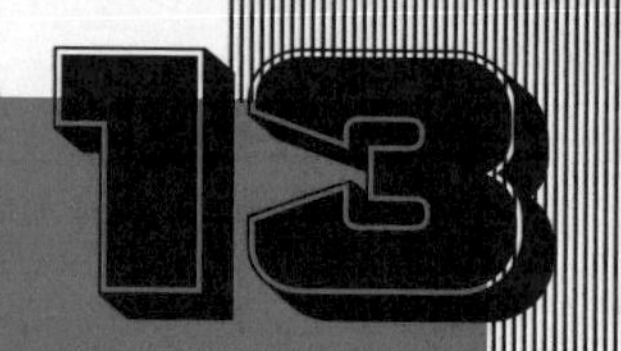

NOVEMBER 8.
LATE AFTERNOON.

Anna Dise clutches a garden hose, soaking the dirt and vegetation around the house she shares with her father. She casts a worried look up at the sky. She's seen this color before—an eerie gray that grows darker toward the east. Not a rainstorm. If only. No, that's smoke from an approaching wildfire. She glances over at her father, Gordy, who's clearing out some underbrush with a weed trimmer.

For twenty-five-year-old Anna and her father, the threat of a blaze is not particularly new. Here in Butte County, in the dry foothills of California's Sierra Nevada mountains, wildfires pose an ever-present danger. Not long ago they heard that the fire was bearing down on the town of Paradise, just a few miles to the east. If it's moving as fast as they think, they don't have long to prepare their defenses.

Gordy, who's in his sixties, has a checklist. The first step is to create a buffer around the home with enough space so the flames can't burn a path to the house. Anna glances up at the only home she's ever known. It's her dad's pride and joy. He's lived here since the early 1990s, before his daughter was even born. She knows he won't let it burn—at least not without a fight.

Suddenly it's quiet, except for the spray of the hose. Gordy has killed the trimmer and is walking toward her. He points toward a patch of earth.

"You missed a spot," he says, a wry smile crossing his face. Anna chuckles and swings the hose, soaking the spot where he's pointing.

Gordy claps her on the back. "Nice work!" he says. He looks up at the sky and Anna thinks she sees a flicker of worry cross his face. But it's gone as quickly as it appeared. "We'll get through this one," he says, and returns to his weed trimmer.

Anna smiles. She and her dad may disagree on just about everything—from politics to motorcycle helmet laws—but on this they agree. This is their home, and they're going to defend it.

But she can't shake a suspicion that this fire is different.

Almost like he's reading her mind, Gordy turns to his daughter. "We're staying," he says, and turns back to the task at hand. Anna sighs. He wants to stay and protect the property, so that's what's going to happen.

Besides, fear is just a feeling, she thinks. *I can ignore it.*

* * *

They couldn't know it at the time, but this would be the deadliest wildfire in recent California history. Dubbed the Camp Fire by officials, it was sparked by a power line in the foothills of the Sierra Nevada mountain range, about a hundred and fifty miles north of San Francisco. The dry and arid region is prone to wildfires, but high winds and a years-long drought combined to make this blaze insatiable. In just three hours, it grew to cover five thousand acres and devoured the equivalent of one football field every three seconds. And in its sights was the town of Paradise, home to twenty-seven thousand people.

Paradise sits on a narrow mountain ridge. When the order to evacuate came, the citizens, mostly young families and retirees, clogged the only four roads that led out of town. A drive that normally took fifteen minutes would take several hours, even as the fire closed in.

* * *

Paradise police officer Rob Nichols goes house to house, rapping on doors and urging residents to evacuate. Nichols is a former firefighter, so he knows how unpredictable a wildfire can be. All the signs he's picking up now—the wind, the smoke, the smell—are telling him the fire that started miles away sometime this morning is going to reach town soon.

His radio crackles. Someone's called in another spot fire—a smaller fire ignited by floating embers from the main blaze—this one across town. This is the fourth call for a spot fire this morning. The last call had been an eighty-foot ponderosa pine tree, engulfed in flames. Nichols and his rookie partner were armed only with a fire extinguisher. *Hardly a spot fire*, he thinks ruefully.

Nichols wipes the sweat from his brow. He calls to his partner two houses down.

"Hey, Kyle, we got another spot fire," he yells. "Let's go."

The two men trot back to their cruiser. All the town's firefighters are already battling the blaze, trying to keep it from reaching Paradise. So putting out the spot fires falls to cops like Nichols and his partner. But as they round a corner, it's obvious they're not going anywhere. The streets are jammed with cars as residents attempt to flee. There's nowhere for them to pull over to allow Nichols's patrol car to pass. Shaken, Nichols gets out his cell phone and dials his wife.

"You gotta take the kids and go," he says.

"We're not in an evacuation zone, though," she replies, her voice quivering.

"You will be soon. Go now," he answers.

Just moments later, Nichols's radio chirps again. It's official: The whole town, including hospitals and schools, is being ordered to evacuate.

Nichols eyes the clog of traffic, eerie beneath the glow of a burning orange sky. The roads need to open up soon, or the residents of Paradise—including his wife and children—won't stand a chance.

* * *

At Ponderosa Elementary School in Paradise, the evacuation can't wait any longer. Parents are picking up their kids but not quickly enough. Twenty-two children need to leave immediately. Kevin McKay drives his school bus up to the front entrance and pulls the lever that opens the vehicle's door. Frightened students, some as young as five, hurry onto

Bus 963 and plunk down on the pockmarked vinyl seats. Two teachers climb onboard too to serve as chaperones. McKay glances into his mirror and can see the frightened faces. It's up to him now.

The forty-one-year-old McKay has been told to drive thirty miles to a school in another town that's out of the path of the fire. There the children can wait until their parents pick them up. But as he pulls into traffic, he feels a surge of fear. The fire is already here. Paradise is burning. Houses, trees, and cars are consumed by flames—the smoke is so thick that McKay's throat is raw. Coughs and sobbing drift up from the back of the bus. It's so dark, a child asks if it's night or day. One of the teachers weeps while on the phone with her fiancé.

There are three main components to smoke, each equally lethal. First are particles, also called "particulate matter," which are bits of burned or unburned substances so fine they slip by your respiratory system's filters and become embedded in lung tissue. "Particles can cause irritation to the lung tissue," says Deepak Sachdeva, an emergency physician. "Inhalation injuries will make it harder to breathe, and you can't absorb the oxygen you're trying to breathe in efficiently, and then you get a lack of oxygen. It can happen slowly or quickly."

Second are vapors, droplets of liquid that can be poisonous if inhaled or absorbed through the skin; these are a dangerous hazard during wildfires.

Last, toxic gasses such as carbon monoxide are common causes of injury. They replace the oxygen in your blood, and when this happens, you'll get a headache and experience reduced alertness; in severe cases it can cause disorientation and coma, as well as irregular heart rhythms, says Sachdeva. These can all lead to death. "Without enough oxygen, heart tissue is damaged too," Sachdeva explains. "This can be reparable if you're able to get out of the smoke quickly. If it's severe, then you can suffer permanent damage to the heart." Some other gasses, like carbon dioxide, aren't toxic to humans, but they occupy valuable space that oxygen needs. Even if treated, carbon-monoxide poisoning can cause "long-term neurocognitive issues, such as irritability, depression, memory loss, and difficulty with focusing and concentrating," according to Sachdeva. "

There's so much congestion, the roads leading out of town resemble parking lots. The passing seconds feel like an eternity as flames creep closer on both sides of the bus. McKay doesn't yell or lay on the horn—he turns on the interior lights, illuminating Bus 963's precious cargo. Maybe if people see his full bus, they'll make space. Finally, an opening appears. McKay inches the bus forward, but he's instantly cut off by an RV. He slams on the brakes, and the bus jerks to a stop.

It takes three hours for the bus to travel just two miles.

Around noon, McKay's bus rolls up to a police officer who is struggling to hasten the exodus. McKay asks about various escape routes, but the cop says the fire and traffic have left only one viable option: Roe Road. It's a street McKay dreads under these circumstances: It's narrow, lined with towering pine trees and ample brush. The ideal fire corridor. And now he's forced to take it.

McKay is parched. He hasn't had water in hours, and he's sweating profusely from the heat. The interior of the bus has surpassed a hundred degrees, and the kids aren't faring well. Some have removed their shirts in a bid to cool down,

while others complain about being unable to breathe. McKay shoves the bus into park and tugs off his polo shirt, then his undershirt, which he hands to a very confused teacher.

> Common materials emit toxic gasses when burning too. Burning vinyl seats, as those found on Bus 963, generate phosgene gas. At high concentrations, phosgene can cause pulmonary edema and death. "Pulmonary edema is when your lung tissue fills with fluid," says Sachdeva. "It's an inflammatory reaction to the irritants, and then fluid builds up." Within minutes, you can effectively drown.

"Make some breathing filters. Rip that into twenty-five squares. Then douse 'em with water and hand them out to the kids," he instructs. She complies, and soon his rearview mirror is dotted with his tiny charges, huffing through torn shirt scraps.

* * *

Officer Rob Nichols is directing vehicles through a downtown intersection when he sees a wall of flame at least eighty feet tall. It's ripping toward him—and toward a gas station, a propane station, and a sporting goods store that's stocked with ammunition.

Unless he acts, Nichols and the people stuck in this intersection are about to get blown to pieces.

Nichols looks around. Across the street, a mini-mall's parking lot is set back from the road, and the wind doesn't appear to be blowing embers that way. He makes his way down the line of cars, banging on windows and shouting for everyone to follow him. More than a hundred residents comply, abandoning their vehicles, carrying pets, unwieldy luggage, and prized belongings. Some weep; others are too dazed to make any sound at all.

Nichols and his rookie partner, Kyle, herd the group into the center of the pavement, away from combustibles. When the first propane tanks explode up the street, people sob and huddle closer. They're too exposed here, Nichols realizes. He glances toward the shopping mall. It's newer, built up to modern fire-safety standards, so Nichols rushes to a coffee shop and smashes the glass door, then unlocks it. The panicked crowd streams in behind him.

As the last of the people enter the shop, a spray of smoldering embers whooshes in through the open door. The crowd leaps to stamp them out. Nichols knows they're still not safe; this fire is burning so hot that even this recently constructed coffee shop could catch, or they could all asphyxiate if the fire passes directly over them.

But for now, the coffee shop is their best chance.

If the air is hot enough, a single breath can be fatal. "If you have a severe thermal injury, it causes so much lung damage that you can't breathe oxygen in," says Sachdeva. "And you'll die."

Inhaling smoke can be incredibly painful. The hot gasses burn your respiratory tract. The superheated air itself can be devastating; immense heat can singe your lungs. "I'd imagine

there's some discomfort, and the discomfort of drowning, if you can't get enough oxygen in," says Sachdeva. "If you start getting angina-type symptoms, it'll feel like a heart attack." 

* * *

Roe Road, where Bus 963 and its twenty-five occupants are stuck, is burning. The sky's no longer inky black from smoke; it's bright red as flames draw nearer. Kevin McKay, his eyes fixed on the path ahead, hears the sobs of a small girl behind him, then her voice.

"There's a deer on fire. I saw it," she stammers. This revelation triggers a fresh wave of panic among other children.

Sparks and burning debris rain down on the bus. The inside is an oven, and its passengers are getting cooked. Sweat drips off of McKay's body as he continues to creep the bus down the road, the path still choked with cars. They don't have long before the flames reach the bus. If the fire incinerates the tires, it's game over. McKay needs to clear a final intersection, and the inferno that is Roe Road will be behind them.

Except no one will let the bus merge.

Suddenly, a pickup truck muscles its way into the intersection. It's blocking traffic, allowing the bus to turn. As the bus lumbers by, one of the teachers recognizes the driver of the pickup; their savior is the school's principal.

The gap between Bus 963 and the wildfire increases. Traffic on this road is moving. It's less than 30 mph, but compared to the snail's pace for the past six hours, it feels like they're going ninety.

The interior of the bus cools, and within a few miles, McKay and the children see something that makes the bus erupt with joy: blue sky. A short while later, McKay pulls into a school parking lot in Biggs, California. As the last of the kids hop off, McKay studies the bus. It stinks of smoke and body odor. Soot and ash coat the inside, and some windows are so charred, McKay can't see through. He pats the side of his battle wagon as he descends the steps.

"You did real good," he says.

* * *

Back in Paradise, Officer Rob Nichols's boss radios him: The fire's still burning, but it's believed some roads are passable. Nichols should find those escape paths and lead the residents in the coffee shop to safety; officials want the town clear by nightfall. Leaving the residents in the hands of other first responders, he hops into his car and starts searching for a clear exit out of town. As he drives, a text comes in from his wife: She and the kids are safe.

The core of the wildfire has passed, and Nichols struggles to get his bearings. Everything is a charred husk of what it was—fast-food restaurants, the local diner, the gas station. Spot fires still burn, and some structures and vehicles are still blazing. But the worst is behind them. The fire has moved on, seeking new fuel.

> Wildfires, particularly ones as immense as the Camp Fire, can also kill by eating up available oxygen. We need 21 percent oxygen to breathe normally. If a wildfire drops the oxygen level to 17 percent, you experience disorientation and poor coordination. Down to 12 percent, and you're tired, nauseous, and dizzy. You lose consciousness at 9 percent, and you die when oxygen levels dip to 6 percent.

* * *

Flames race down the canyon walls that tower above Anna Dise's home. She dashes across her front yard toward her car, clutching a stack of books in one hand, her laptop in the other. Her two dogs, Luna and Sirius, are on her heels, barking in alarm. Across her lawn, a row of trees is burning wildly. Thick smoke is everywhere. Sporadic loud cracks hang in the air as burned tree limbs fall to the ground, sending showers of embers airborne.

She looks back to her house. Her father's still in there, and they're out of time. Flying embers and choking smoke have turned the world an apocalyptic dark red. Anna and the dogs clamber into the car.

> If you're trapped in a house being engulfed by a wildfire, you have anywhere between two and ten minutes before smoke inhalation renders you unconscious, then kills you. "Outside, you can survive moderate smoke inhalation for a prolonged

period of time, but there can still be permanent damage to your body," says Sachdeva. Inside a house, the smoke is more intense and dangerous. "Your lungs will be directly affected by the smoke and heat," explains Sachdeva. "But any organ in the body will suffer damage without oxygen, including your heart, kidneys, and liver. Brain damage would be a big issue too. Damaged brain tissue is not reversible," says Sachdeva.

She lays on the horn and shouts "Dad!" out the open window. She can feel the heat scalding her face. *What the hell is he doing?*

At last, Gordy bursts from the home, a pile of his beloved vintage motorcycle T-shirts in his arms. Floating ash has grayed his hair, and Anna looks past him to see flames popping up between the slats of their wooden deck.

He runs to the car and tosses the shirts inside. But instead of jumping in, he tells his daughter he'll be right back.

"Dad! No, we have to leave!" she pleads.

But it's too late. Gordy's already dashing back onto the porch and sidestepping the flames. He disappears through the front door just as a white-hot wall of fire envelops one side of the house. Anna can only honk and scream again as the flames begin to consume the home's roof. Her eyes lock on the front door, desperate for her father to emerge. Praying that he will.

A sickening pop from the house cuts through the noise of the blaze. One wall teeters, then collapses, sending up a shower of burning debris, some of which land inches from the car. The fire rips across the roof, flames now visible through the windows.

Anna is hoarse from the smoke and from her screams. Her chest is tight with fear. Another violent sound, this time to her left. She turns to see the car parked beside hers erupting in flames with a violent whoosh. She has seconds before the fire reaches her own car.

The dogs whine frantically as Anna squints at her house. It's engulfed. Chunks of the roof are collapsing, columns of fire shooting high into the sky. *Oh my God, there's no way Dad survived*, she thinks. She slams the car into reverse and floors the accelerator.

But the car doesn't move. The wheels spin madly. She cracks the door and sees why—her tires are melting off the rims, pooling in sticky puddles of black.

She flings the door open and jumps out, Luna and Sirius following. The trio sprint down the driveway. In her periphery, Anna sees the fire closing in on all sides. Behind her, the car they escaped explodes in a giant fireball.

There's a soggy ditch in front of her neighbor's property. She dives into it, yanking the dogs down with her. She slathers herself and the dogs with muddy water. All three tremble. Flaming leaves drop onto them. Embers from burning pine cones swirl by, stinging Anna's face. A tree slams to the ground nearby, startling her. Then a neighbor's propane tank explodes. Sirius whines, eyes wide. Luna shakes uncontrollably with fear.

Night is falling, but the sky remains volcano red. Anna wonders how long she can last when another tree cracks apart and crashes to the ground. Sirius bolts and races toward the house, still ablaze. Anna cries out and stands to give chase, but a mess of burning embers prevents her from moving. The ditch is her only refuge from the conflagration.

As the dog disappears into the glow, Anna sinks back into the ditch and clutches Luna, her face buried in the dog's muddy, smoky fur. Heavy sobs rack her body.

First her father, now her dog.

And still the fire rages—igniting more propane tanks, cars, dozens of trees. Hours pass, and she holds tightly to Luna.

When dawn arrives, sunbeams slice through the smoky air. Spot fires lap at any available fuel, trees glow red, but the main part of the fire has finally passed. Stiff and parched,

Anna and Luna stagger from the ditch, caked with mud and ash.

Moving toward her house, Anna sees it's nothing but a pile of charred beams. Her eyes fall on the metal carcass of her melted car. There, on a small patch of charred fabric where the backseat used to be, lies a piece of one of her father's T-shirts.

Grief overcomes her and it's hard to breathe. In between sobs, she hears Luna bark. The dog won't stop barking at something on the ground. The object is moving. Breathing. Anna races over. "Sirius!"

She collapses on the ground next to the ash-covered dog, gently shaking him. It takes a moment, but he rouses. She flings her arms around his neck and he licks her streaming tears.

* * *

It took two weeks for the Camp Fire to be 100 percent contained. Before it was finally extinguished, the wildfire torched more than 150,000 acres and 18,000 structures. Eighty-five people died, making it the deadliest fire in the United States since 1918.

The origin of the fire was traced to an aging, poorly maintained power line. The line's owner, Pacific Gas and Electric, eventually pleaded guilty to eighty-four counts of involuntary manslaughter. The utility paid $3.5 million in fines and $13.5 billion to victims and families of the dead.

Only a third of Paradise's residents, some nine thousand people, opted to return. Rob Nichols was among them and continued to patrol the town. Kevin McKay resumed work as a bus driver for Paradise Unified School District, though the Ponderosa Elementary School from which he saved the children didn't reopen. Anna Dise moved to nearby Chico, California, with her dogs.

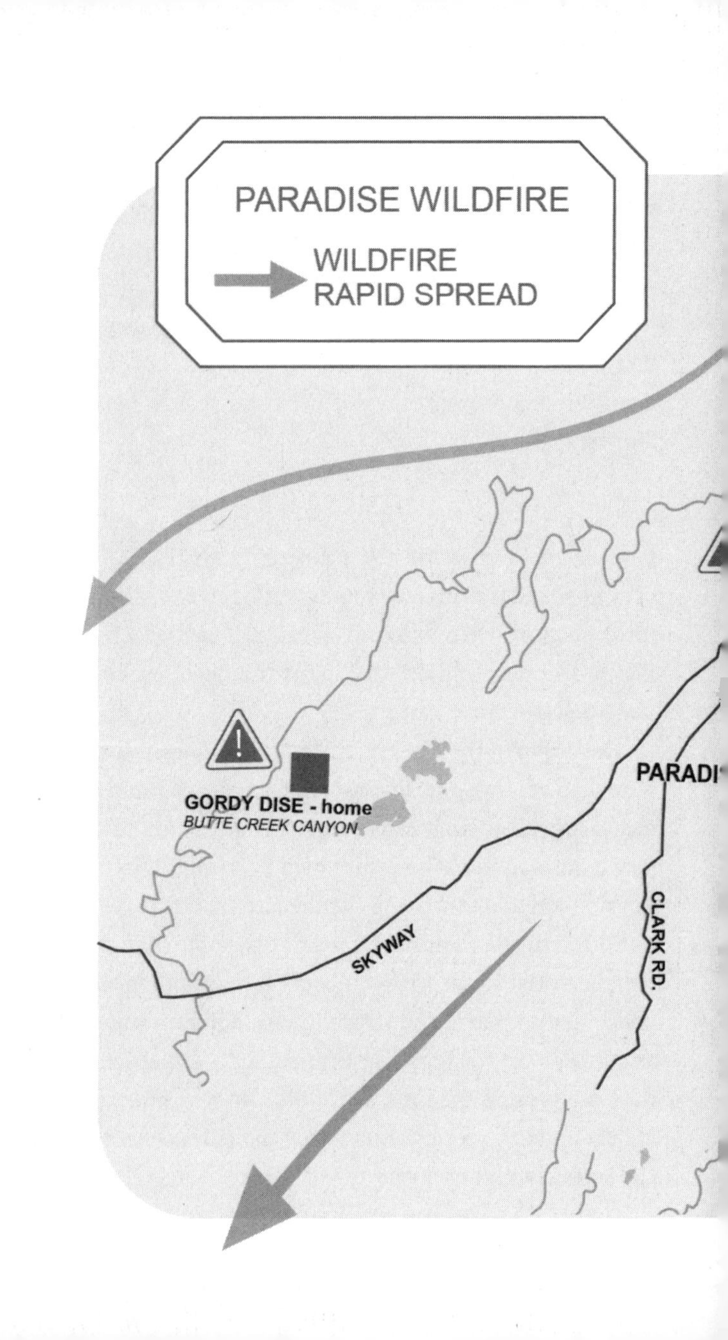
PARADISE WILDFIRE
WILDFIRE
RAPID SPREAD
GORDY DISE - home
BUTTE CREEK CANYON
SKYWAY
CLARK RD.

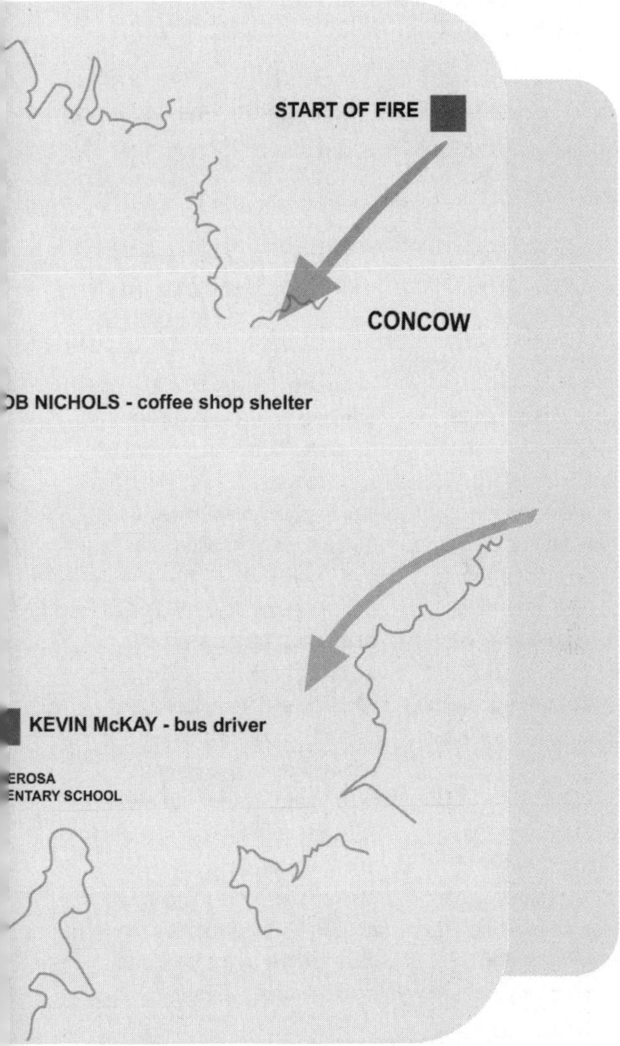

START OF FIRE

CONCOW

OB NICHOLS - coffee shop shelter

KEVIN McKAY - bus driver

EROSA
ENTARY SCHOOL

HOW TO SURVIVE A WILDFIRE

Your wildfire survival checklist depends on your location, but it's always important to consider the following:

REMAIN CALM. Otherwise, you can't properly analyze the situation and make good, rational choices.

PLAN TWO ESCAPE ROUTES AND MEETING LOCATIONS. If you live in a remote area, assume one route may be blocked, either by fire, downed trees, power lines, or a throng of escapees. Predetermine where you'll meet loved ones after leaving the area too. Cell phone towers may be down, so real-time communication can be hard.

ACT IMMEDIATELY. Wildfires can move faster than a freight train, far quicker than you can run. If you see smoke and embers intensifying, don't hesitate—mobilize immediately.

IF YOU'RE TRAPPED AND CAN'T EVACUATE, CALL 911. If you're trapped at home, fill all sinks and tubs with cold water to have extra water ready for drinking or throwing on errant embers. Close—but do not lock—doors and windows. Stay inside your house, away from outside walls or windows.

IF YOU'RE IN YOUR VEHICLE AND ABLE TO DRIVE, GO SLOWLY. Make sure your hazards and headlights are on, roll up your windows, and set your air conditioner to recirculate. Cover your mouth with a respirator mask or other dry material. Contrary to Kevin McKay's well-meaning actions on Bus 963, wet fabrics may react poorly in the heat and instead create steam.

IF DRIVING ISN'T POSSIBLE, PARK IN A FIRE-RESISTANT AREA. Avoid vegetation of any kind. Choose a paved lot over a field, even better if there's a solid wall to hide behind. Call 911 and give your location. Don't shut your car off. Shut all vents, roll up the windows, and leave the engine on. You need to be able to drive away at a moment's notice.

GET AS LOW AS POSSIBLE. Situate yourself at least below your windows, ideally on the floor of your home or vehicle. Cover yourself with a blanket, anything that's not made of synthetic material, which can be more prone to melting. If you're in your car, your tires may explode from the heat and fire nearby; don't freak out when you hear them go. Despite what Anna Dise did, it's safest to stay in your car and wait for the flames to pass.

IF YOU'RE ON FOOT, STAY UPWIND AND DOWNHILL. Assess which way the smoke is blowing and move in the opposite direction if you can. Wildfires typically advance uphill, so remain below the fire whenever possible. Get to a clearing free of fuel—no vegetation, away from propane tanks and flammable structures—and find a depression or ditch in which to lie. If you can get near or into water, even better. If not, lie face down, dig a hole, and stick your nose and mouth inside. Try to cover your body with dirt, mud, or anything nonflammable. Shed any clothes that are made of synthetics. Call 911 and give your location.

IF A FIREPROOF STRUCTURE IS NEARBY, GO THERE. Officer Rob Nichols knew the Paradise coffee shop was new enough to have met fireproofing standards. You should be sure to know several similar places around your area and do your best to make it to one.

SUCKED INTO A TORNADO

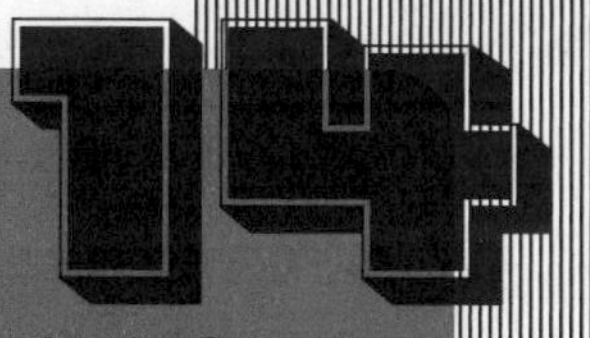

OCTOBER 20. EVENING.

Chris Tuveng is hurrying to make it to Little Caesars. He decided to dash out during halftime of a football game he'd been watching with his family. It's not a long trip, so he should have plenty of time to grab a pizza and get back for the second half.

He glances up through his windshield at the evening sky and can make out dark thunderheads billowing on the horizon. Thunderstorms started in the Dallas, Texas, area late that afternoon, and a tornado watch alert was issued about forty-five minutes ago, around eight o'clock. Just before Tuveng left his house, the football broadcast was interrupted by local weather reports, warning that a nasty storm cell would pass through at about nine fifteen. Still, he thinks he'll be home well before then.

Pulling into Marsh Lane Plaza, a generic strip mall not far from his home, the fifty-three-year-old Tuveng parks and heads into Little Caesars. The place is unusually crowded, and the store is out of pre-made pizzas. Tuveng places his order and takes a seat near the door to wait.

Suddenly the front door blows open, so he stands up and pulls it shut. He's heading back to his seat when the door flies open again. *That's odd*, he thinks, and goes to close it. Reaching for the handle, he glances up at the patch of sky visible beyond the breezeway overhang. There's an ominous swirling mass in the air—dark even against the twilight of the sky behind it. It looks like an enormous cloud of debris and dust. A low rumble grows into a deafening roar that drowns out all other sounds as it bears down on him. First the cloud is across the street, but when he blinks again, it's in their parking lot. Then he realizes what he's seeing.

"Tornado!" Tuveng shouts, spinning around to the customers crowded into the pizzeria. He barely sees workers and guests scramble to the back of the store, ducking for cover, when his feet leave the ground and he's sucked out the open door.

He's airborne for a split second before being flung to the ground. His body slams the concrete, but before he can get to his feet, the tornado scoops him up again and flings him farther down the breezeway. He lands, blinking hard against winds that are gusting up to 140 mph.

> In violent tornadoes—such as EF3s to EF5s, where winds can range anywhere from 140 mph to 300 mph—flying debris is your biggest threat. These tornados can carry cars for considerable distances, rip entire homes off their foundations, and turn metal, glass, and other building materials into lethal projectiles. Tornadoes can span an area more than a mile wide and stay on the ground for more than ten minutes.

Debris whips through the air, pelting his body. He's wearing just a T-shirt and shorts, and some of the objects—stones, shards of glass—lodge themselves in his exposed skin. But there's nothing he can do as the tornado rips him off the ground again. The wind is determined to take him with it, and the roar is deafening.

> When debris strikes your body, soft tissue damage occurs. You get abrasions—scrapes and cuts—like Chris Tuveng did. Debris can also lodge in your skin. "If something is impaling your body, leave the item inside your body until you can get professional medical attention," says Matt Cummins, a board-certified emergency physician. "You need imaging first, as the projectile can be plugging broken blood vessels. Pull it out and you risk bleeding to death."

He's deposited near a concrete support pillar that holds up the breezeway roof. Instinctively he wraps his arms around the column, squeezing with all his might.

You're going to die here, he thinks. A stream of grim scenarios flash through his mind. Will the tornado blow him so high into the air that he won't survive the landing? Will a huge piece of debris smash into his head, killing him instantly?

> Head trauma is also common. This can happen either by being struck by an airborne object or if you're lifted off the ground and flung around. "You don't have to fall from very high to have a serious injury," says Cummins. "Your neck can snap, or you can have tissue swelling inside your head, increasing pressure on your brain." Severe head injuries are often the cause

of death from a tornado. Objects can pierce your skull too, causing instant death.

Will a large slab of glass stab him, and he'll bleed out? He closes his eyes as the tornado rips his shirt to tatters, and he prays. *Hold on as long as you can.*

Debris can gouge chunks of flesh off. If you're left with a gaping wound like this, first stop the bleeding. "Apply pressure, bandage it, and elevate the wound above the heart," says Cummins. "For missing skin, a wet bandage is better, to prevent tissue drying, but any bandage is better than no bandage."

* * *

Chris Tuveng is caught in the teeth of an EF3 tornado, where winds can reach speeds of 165 mph. This tornado, which touched down sooner than forecasters expected, is more than a half mile in diameter. For thirty minutes it churns a path of devastation over fifteen miles, uprooting trees, leveling houses, tossing vehicles around like toys, and sucking people right out of buildings.

And directly in its sights is the Marsh Lane Plaza, home to a Little Caesars pizza shop and a Planet Fitness gym.

Rachel Ellis-Piantanida gets the alert about the storm on her phone. The twenty-six-year-old works at the twenty-four-hour Planet Fitness. Tonight, it's just her, a handful of other employees, and only one gym member. Tornado alerts are new to Ellis-Piantanida; she recently moved to Dallas from California. She looks outside. It's raining madly, coating the gym's giant windows in sheets of water. She arrived for her shift only fifteen minutes ago and is now wondering if she remembered to close the windows in her car.

Without warning, the gym's TVs go black. Looking around, her coworkers are equally confused. Then the power cuts out. Then a loud crash.

All the windows shatter as glass explodes everywhere.

Rain blows sideways into the building. The noise is overwhelming, so loud it vibrates her body. Ellis-Piantanida screams and rushes toward the gym's locker rooms, on the heels of her coworkers and the gym member. She's halfway there when there's an enormous crack. A beam falls from the ceiling behind her, sending a cloud of debris through the air.

The gym is disintegrating.

"Get to the showers," someone shouts.

The group keeps running toward the locker rooms. Another deafening bang. The huge reception desk has been upended by the tornado, and it's skidding across the floor of the gym. It slams into treadmills and ellipticals, battering them like a hammer. Frantically, Ellis-Piantanida and the others make it into a locker room and head to the showers.

Except it's not safer here.

The tornado is now bearing down on them from above, tearing the roof off the building. Ceiling tiles crash down on the shower floor. Ellis-Piantanida tugs the group toward the sinks. They can huddle beneath them. But the noise is sickening—creaking metal, howling wind, clattering debris. It's as if the tornado is searching for them.

* * *

Chris Tuveng's arms are wrapped around the support column as the tornado assaults his body. He's managed to make it to his feet, his chest pressed against the column. The wind is pushing him into the concrete with such force, it feels like he's getting crushed.

> If you're stuck, pressed against something, as Chris Tuveng was with the concrete pillar, the force of the wind will feel crushing. At speeds above 130 mph, you won't be able to speak, and you may be incapable of taking a breath if the wind is coming at your face. "Breathing works when your diaphragm goes down and the negative pressure inside your lungs pulls in air," says Cummins. "If the air pressure outside is so low it's causing 130 mph winds, it'll be very hard to breathe. You won't suffocate, but it will feel weird." There's still hope, though, because the human body can withstand wind speeds of up to an impressive 450 mph.

He's getting hit now with even bigger pieces of debris—branches, trash bins, rocks. Around him, the Marsh Lane Plaza is rapidly being reduced to rubble. The winds are peeling the storefronts away. Tuveng sees a dumpster whiz by and smash through the dining room of the Little Caesars. A religious man, Tuveng continues praying that the onslaught ends soon. Or that, if he is going to die, it will be painless.

It feels like he's been holding this pillar for an eternity, even though it's only been ten seconds. But the force of the storm is too intense. A ferocious gale pries him off the pillar. He bounces off the ground, his outstretched arms grasping for anything to grab on to. But there's nothing. He's lifted into the air again, then slammed down. Once. Twice. Each time he's sucked into the vortex, he braces for the inevitable landing, certain that the next impact will kill him. Several vicious thuds later, the tornado hurls him onto the hood of a truck, in front of the battered pizza shop.

In 30 percent of tornado injuries, broken limbs occur. "Open fractures—where bone pokes through skin—are the most problematic because then bone can get infected," says Cummins. "Bone can easily break after being hit with a projectile flying around at extreme speed."

Blood pours into his eyes from cuts on his head and face. The tornado is still raging and he knows it's a matter of seconds before he's tossed into the air again. There's little to hold on to on the hood of this truck, so he rolls over and drops to the pavement.

Down here, Tuveng wraps his arms around the truck's tire. It's his best chance for an anchor point. He feels the tornado tugging at him, his legs lifting off the ground. He grips the tire tighter. Behind him, a smaller pickup truck flips on its side and skitters across the parking lot. He hears the eerie sound of metal scratching across the pavement, then a sickening crunch as the truck smacks into a parked car. *Please don't blow this truck away too.*

Tuveng's arms and hands ache as he struggles to maintain his bear hug on the tire. But his strength is giving out.

Then, as suddenly as it arrived, the tornado passes. Tuveng's world goes from utter chaos to complete stillness in the blink of an eye. Power gone, everything black, and the only sounds he hears are chirping car alarms and screams from amid the rubble. It looks like a war zone.

He staggers to his feet. Looking down, his legs are black from debris and streaked with blood. He catches a glimpse of his reflection in the chrome trim of the truck that helped save his life; his face is unrecognizable, a mess of blood and dirt.

But he's alive.

* * *

Chris Tuveng miraculously escaped serious injury. No bones were broken and he sustained only superficial cuts and scrapes across his back, legs, and head. He estimates he was inside the tornado for a total of thirty seconds, though it felt like a lifetime. He was treated in the hospital and released after a few hours, sore and bruised, but grateful to walk away. Small bits of debris remain lodged in Tuveng's body; doctors were unable to remove them all. Workers and other customers inside the Little Caesars store also suffered scrapes and cuts, though none were seriously injured.

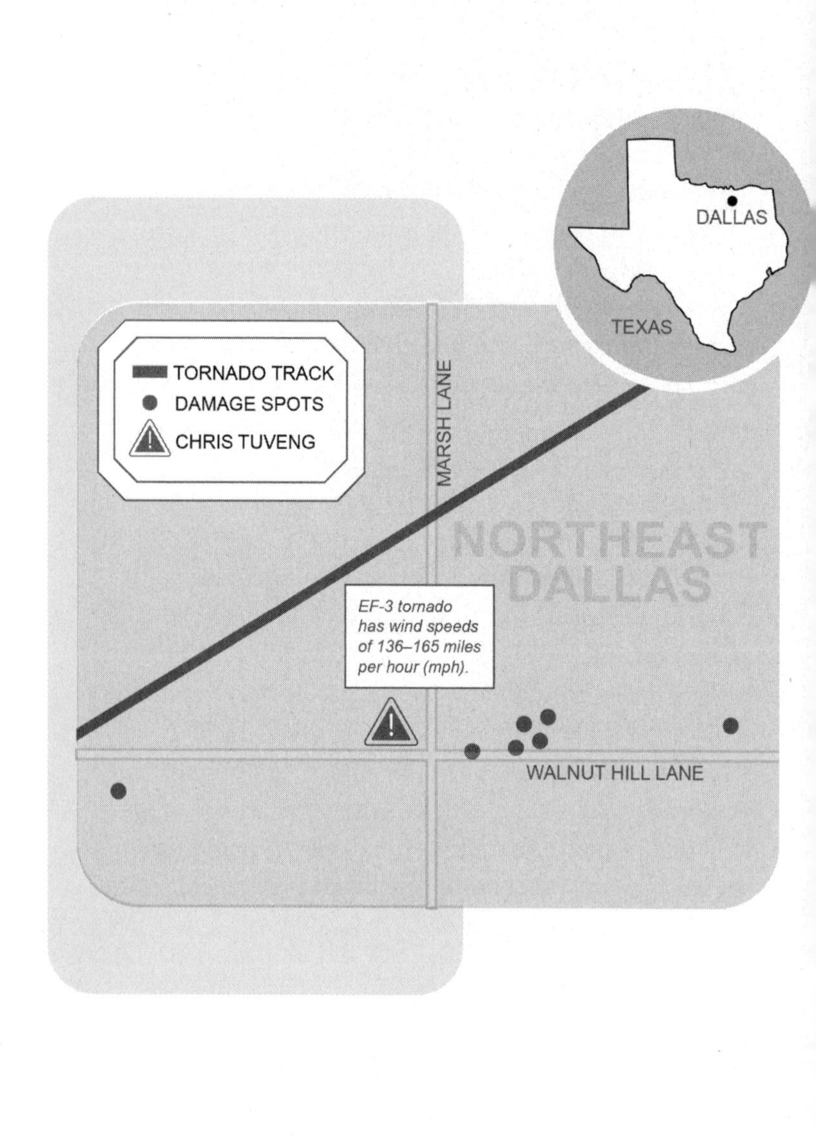
DALLAS
TEXAS
TORNADO TRACK
DAMAGE SPOTS
CHRIS TUVENG
MARSH LANE
NORTHEAST
DALLAS
EF-3 tornado
has wind speeds
of 136–165 miles
per hour (mph).
WALNUT HILL LANE

> Doctors may not be able to remove all the tiny particles stuck in your body. Chris Tuveng believes some of the tornado's debris may remain inside him for the rest of his life. "We typically leave small debris that doesn't impact functionality in the body," says Cummins. "We can cause more damage when removing it," though he adds that anything that's painful or causing trauma is definitely removed. Over time, any small bits of glass, wood, or metal may work toward the surface of the skin, where they could then be cut out.

Rachel Ellis-Piantanida was unhurt. The space under the sinks protected her and the rest of the group, despite the entire Marsh Lane Plaza being demolished. Only ten people were injured in the tornado. Thankfully, no one died.

HOW TO SURVIVE A TORNADO

There are plenty of ways to prepare your home and body to beat the odds against a tornado. A little planning can go a long way.

FIND SHELTER IMMEDIATELY. Tuveng admits he should've heeded the weather warnings and stayed home. If a warning comes through and you're home, the safest spaces are the interiors of your basement. No basement? Any interior room works, so long as you're away from windows and ideally on the ground floor. Closets, center hallways, and bathrooms work.

ONLY GET INTO METAL BATHTUBS. Waiting out a tornado in your bathtub may sound like a good idea, but that's not true for tubs made of plastic or fiberglass. Those are easily pierced by flying debris, and you could be seriously injured.

REMOVE POTENTIAL PROJECTILES. In the space in which you're sheltering, get rid of anything that could be harmful if it were to go airborne. This includes things like clocks, artwork, lamps, and smaller pieces of furniture.

KNOW WHAT'S ABOVE YOU. Heavy objects like pianos, refrigerators, or dressers can fall through upper floors if the tornado hits your house. Make sure wherever you're sheltering doesn't have anything sizable over your head.

PROTECT YOUR BODY. Drag a mattress over yourself, if possible. Thick blankets or bedding work too. Lie face down and cover your head, either with a bike or ski helmet or your hands.

AVOID COMMERCIAL BUILDINGS. Bigger, long-span buildings, like shopping plazas and gyms or malls, are among the worst places to be when a tornado hits. They're constructed to be supported by outside walls, and they collapse quickly during a tornado, as Rachel Ellis-Piantanida experienced in Planet Fitness. If stuck in such a location, get under a door frame or a structure that can protect you from falling debris, as Ellis-Piantanida did by going underneath the sink.

FIND ANCHOR POINTS. If sucked into the storm, take a page from Tuveng's book and hold on to anything that doesn't seem like it could be blown away. You'll stand the best chance of surviving.

LIE IN A DITCH. If caught outside without anything to hold on to, your only bet is to find a depression in the ground and lie in it, face down, with your arms covering your head and neck. Do not lie on asphalt; if the winds are strong enough, the tornado can rip up the pavement.

SPLINT BROKEN LIMBS. If something breaks, Cummins says the pain may be excruciating, but get the limb straight enough to ensure good blood flow. You want to feel a pulse in the affected area. "The longer you're without blood flow, the worse it is," says Cummins. After the limb is straight, immobilize the broken portion. "Movement equals tissue damage inside," says Cummins. He says to use a tree branch and shoelaces if you have to, but stop the movement of the joint above and below the break. If you can't splint it, make a sling out of a T-shirt. "You want to take the weight off the broken area," says Cummins.

SWEPT AWAY BY A FLASH FLOOD

JULY 15. 3:30 P.M.

The Arizona sun is blistering as Julio "Cesar" Garcia carries his one-year-old daughter, Marina, up the path alongside the East Verde River. Cesar glances back. The remaining fourteen members of his family are taking their time making their way in the steamy heat, but soon they'll be splashing around at the swimming hole known as Water Wheel.

"That water's gonna feel good!" Cesar's sister Maria shouts from the back of the pack. Today's excursion is Maria's idea, in celebration of her twenty-seventh birthday tomorrow. Large family outings are the norm for the Garcias. The family matriarch, Selia, came to Arizona from Michoacán, Mexico, with six children, all of whom settled in the Phoenix area.

Cesar's two brothers and two sisters are here, along with his brother-in-law and five nieces and nephews, aged between two and thirteen. So is Marina's mother, Abigail, and Cesar's son, Acis, eight. Several family members work in restaurants together, and when they aren't working, they gather to let the cousins play. And today the kids are itching to play.

"How much longer?" whines a child from somewhere behind Cesar. He understands; it feels like they've walked at least a mile since they parked their cars and started the trek.

"Soon!" one of the adults soothes. Maria catches up to Cesar as the group crosses the East Verde River and turns south to hike upstream along a tributary called Ellison Creek, which will lead them to Water Wheel.

At this time of year, Ellison Creek is just a trickle, flowing between boulders at the bottom of a narrow box canyon. Red-hewn canyon walls loom on either side of them, sometimes reaching fifty feet or more, dotted with green lichen and topped with cottonwood trees.

"See? I told you this would be perfect." Maria smirks at Cesar, gesturing to the scenery. "And you were worried about rain."

Cesar grins back, rolls his eyes, and shifts his daughter to his other arm. When he checked the forecast earlier, storm clouds were predicted, but Maria's right: This is a beautiful afternoon and an idyllic locale. He'll be happy to get there, put Marina down, and enjoy a cool-down dip.

The group climbs across the smooth rocks that line either side of the creek. They're making good time. Cesar wipes the sweat off his forehead, looking forward to the cool water of the swimming hole. He hopes it's deep enough for the grown-ups to get a full-body soak.

Suddenly, a dark shimmer from above catches Cesar's eye. He freezes for a second. *What the hell is tha—*

When we freeze for a long time, this is actually called tonic immobility, and it can be a response to a threat or stress. Physiologically, tonic immobility can happen when your brain can't properly process that threat, says Anthony Giovanone, a doctor of osteopathy and a psychiatrist. "Your prefrontal cortex isn't engaging, and you're stuck in this limbo, mentally and physically," he adds. Some people are literally unable to move, while others move very slowly. "After facing a large threat—which could be seeing a wall of flash-flood water coming at you—you're so scared that you completely give up," says Giovanone. "Your brain believes nothing's going to work so it tries to conserve energy and hope for the best."

Before he can finish his thought, it registers. A two-foot-tall wave of inky black water, maybe a hundred yards ahead—and it's barreling directly at them. His hair stands on end when he sees what's directly behind it: another even larger wave, at least five feet tall, carrying massive tree limbs with it.

It's a flash flood. And Cesar knows they won't be able to escape it.

"Get to higher ground!" Cesar screams to his family as he turns to run. By the time the words leave his lips, the first wave has closed half the distance. He sees his family frantically trying to climb up the banks of the creek, but there's not enough traction on these slick rocks. They're too polished by centuries of rushing water—and now that rushing water is about to swallow them all.

His arm tightens around Marina as he tries to scramble to safety. A foot to his right, one of Cesar's nephews falls as the two-foot wave slams into them. In seconds, the entire canyon basin is underwater. Cesar reaches down with his free hand to grab his nephew, but just as his fingers close around the boy's shirt, the five-foot wave bulldozes them from behind.

Instantly, his nephew is ripped away. Cesar's feet are swept up, and he and Marina topple into the muddy torrent. *Hold on to the baby*, he thinks as the duo sink beneath the surface. Cesar squeezes his eyes shut. In the dark and muddy floodwater, no ambient light seeps in.

He has no idea how fast they're being shot downstream,

but it feels like he's stuck inside a waterfall. The flood's current is so immense, Cesar feels it prying Marina from his arms. He presses the infant harder against his chest.

If you don't hold her tight, she's going to be gone too.

* * *

The flash flood that enveloped Cesar and the Garcia family was later determined to have a flow force of 1,950 cubic feet per second, around 14,500 gallons. Picture a small swimming pool's worth of water whizzing by every time you blink.

Tropical air from the Gulf Mexico had blown in, turning the arid Arizona desert into a steamy soup. When that wet air hit the Mogollon Rim—a two-hundred-mile-long series of cliffs thousands of feet above Ellison Creek—it concentrated into pockets of thick thunderclouds that dotted the Rim for miles. These clouds merged, creating a monsoon some eighteen miles away from the Garcias. Shortly after one o'clock heavy rain began to fall, sometimes as much as 6.5 inches per hour.

To make matters worse, the monsoon emptied almost directly upon the aftermath of a raging wildfire—the Highline Fire, which had burned furiously for two weeks a month earlier, destroying nearly seventy-two hundred acres just below the Mogollon Rim's crest. This meant there was ample ash and scorched debris for the water to wash away. And fewer trees and vegetation to slow the heavy rain's flow down the Rim's steep face.

At 1:43 p.m., the National Weather Service issued a flash-flood warning. With spotty cell service in Ellison Creek, the alerts didn't reach the Garcias.

As the monsoon rains fell, the flood gained force, racing down the Rim's cliffs. In two hours, it reached Ellison Creek and the Garcia family.

* * *

Underwater, tumbling in the sludgy flood, Cesar's thoughts flash to his family. *Are Abigail and Acis okay? What about all the other children? And my mom?* Then a sharp pain in his legs. Something huge hit him; he's not sure, but it feels like a large rock. Another enormous blow, this time to his ribs. Cesar winces and wraps himself like a ball around tiny Marina, praying nothing strikes her. And that she doesn't drown.

> **If you're floating unprotected in the water, large pieces of debris can slam into you, at astonishing speeds. Tree limbs, boulders, even cars will be moving rapidly. One hit from one of these objects can be fatal. Cesar and Marina were extremely lucky. If you're in a similar situation, "lie on your back with your feet forward as the river takes you," says Matt Cummins, a board-certified emergency physician. "Protect your head with your hands. You're at the mercy of the current, but that position would be the best way to avoid major head and chest trauma."**

Suddenly his head shoots up, clearing the surface. He takes a deep breath of air and sees Marina do the same. He shoots a glance upstream—up*river* now—and figures he's been pushed twenty yards from where the wall of mud and debris first pummeled him. As the flood carries him farther downstream at tremendous speed, he looks down at his tiny daughter. She's shaken and scared.

It's impossible to swim in the flood. The current is too strong, and the tons of debris in the fast-moving water mean there's no space in which to move. With some branches and tree trunks more than forty feet long, he knows it'll be a miracle if he and his daughter aren't crushed to death.

> **If you're in the water during a flash flood, as Cesar was, the current will move extremely fast, you won't be able to stand up, and you may not be able to see which way you need to swim to find safety. If the body is getting pummeled by fast-moving water, your muscles will tire more quickly, you'll need more oxygen to keep going, and it'll be laborious. "Swimming against a mild current is very hard; against a flash flood like this, it'll be impossible," says Cummins. "You'll lose energy and tire within minutes. Your best bet is to go with the flow and keep floating until the flood spreads out."**

As they careen downstream, Cesar spots it: a tree jutting out over the water a few feet ahead. Still gripping Marina close, Cesar frees a hand and thrusts it at the lowest branch.

His fingers connect and close tightly around this lifeline. Father and daughter jerk to a halt, as Cesar strains to maintain his grasp.

> We're constantly told to stay calm during stressful situations, but that's slightly mislabeled advice, according to Giovanone. "You want to keep your prefrontal cortex engaged. Cesar Garcia had a plan; he knew to protect his daughter, he worked out which upcoming branches he could grab. There's a lot of thinking required for all those movements, all from his prefrontal cortex. Really, instead of staying calm, you're trying not to freeze."

But the current is too strong. The force of the flood pries Cesar from the branch, and he and Marina are once again plunged back into the fast-moving muck. Rocks and charred tree limbs inside the churning flood assault Cesar's body. No matter what slams into him, no matter how painful the impact, he refuses to loosen his bear hug on his baby girl.

> The biggest threat is drowning. The average person can hold their breath between thirty and ninety seconds before reflexively breathing. Training and practice can lengthen that time (Navy SEALs can hold their breath for five minutes).

After what feels like an eternity, they surface, another twenty yards down Ellison Creek. Ahead, another branch, about to be within Cesar's reach. *Do not let go this time.* He snags the tree's limb and grips it with every ounce of his strength. His head and chest—and Marina—are clear of the water, though his legs and lower torso remain submerged in the flood. He looks down at his daughter, caked in thick mud, her dark eyes full of confusion and fear. With all the muck, he can't tell for sure, but Marina seems uninjured. She's hardly crying.

Within five minutes, the worst of the debris passes. The last of the burned trees the size of buses rip past, and now

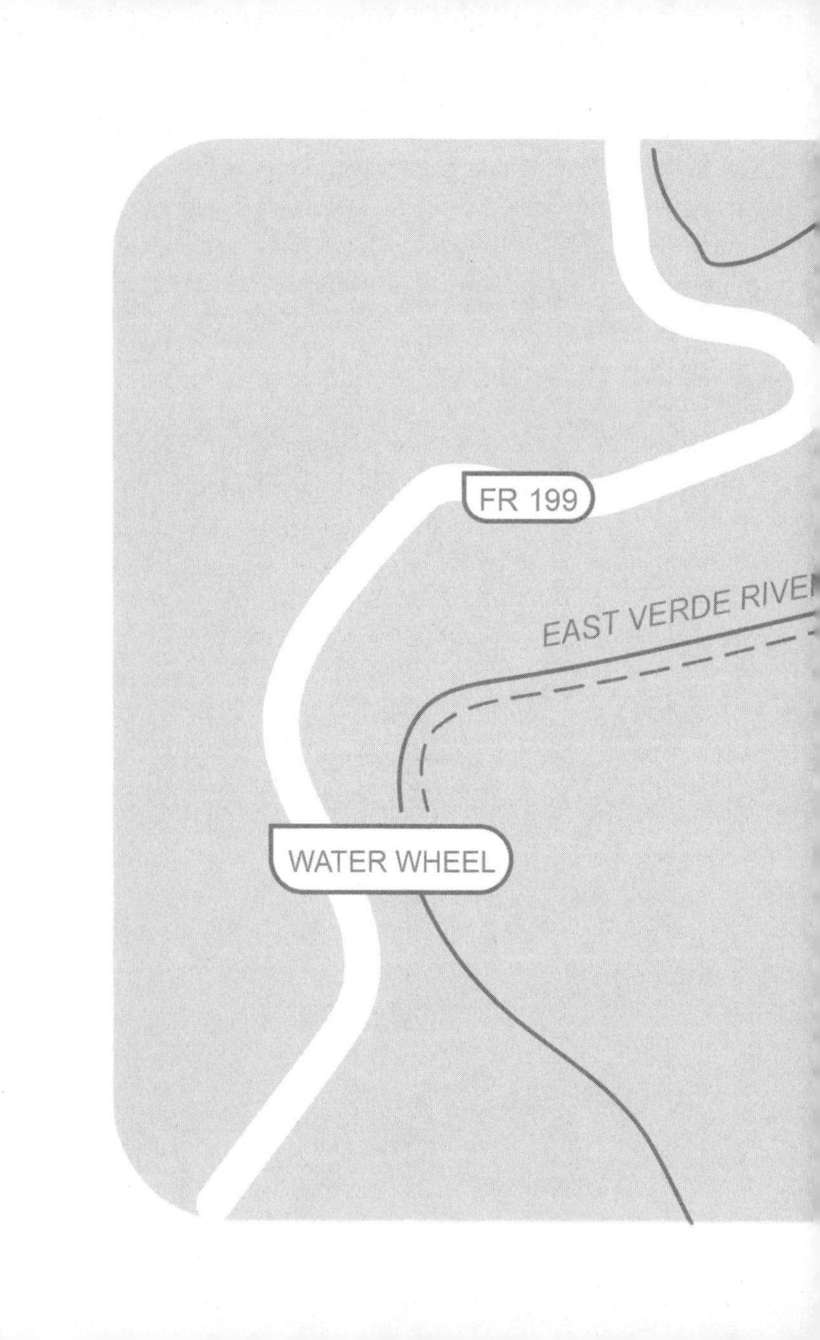
FR 199
EAST VERDE RIVE
WATER WHEEL

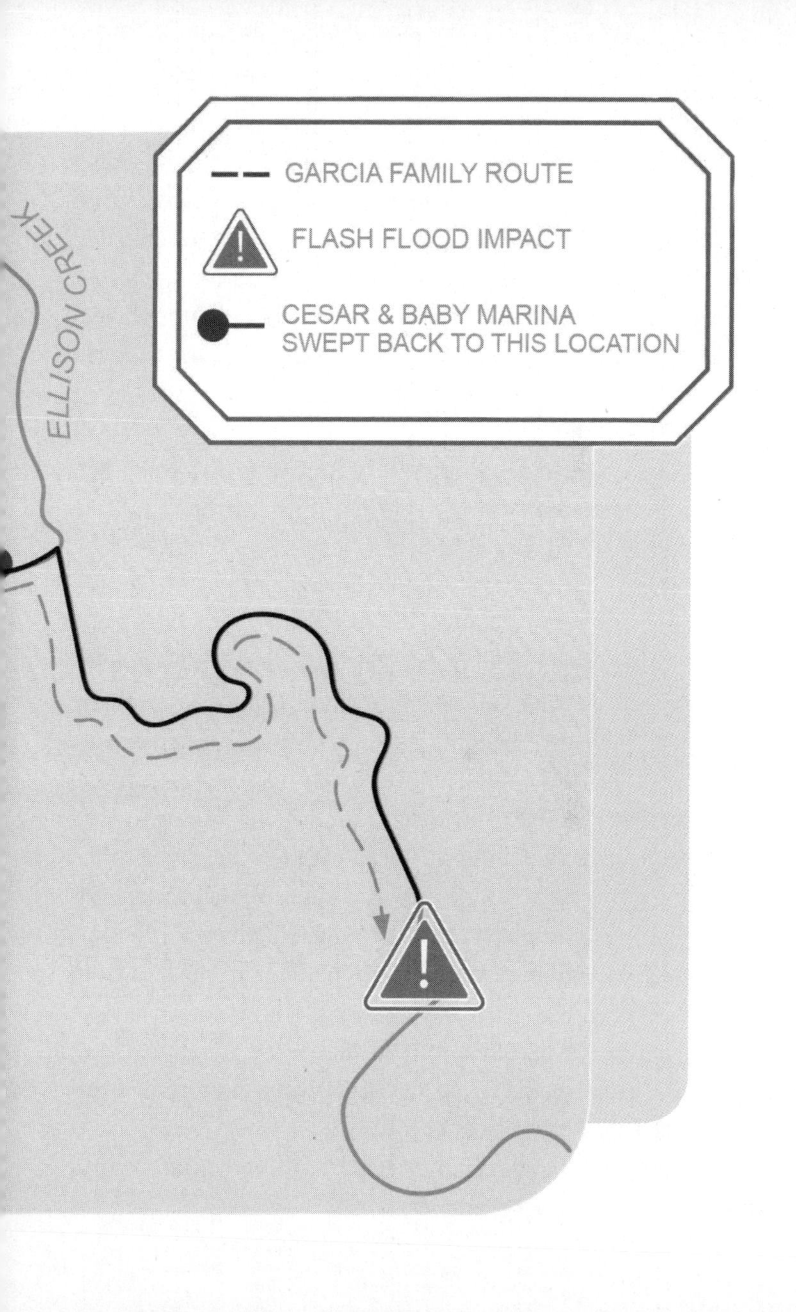
GARCIA FAMILY ROUTE
FLASH FLOOD IMPACT
CESAR & BABY MARINA
SWEPT BACK TO THIS LOCATION
ELLISON CREEK

it's just a fast-moving river, with pockets of rapids. Clinging to the tree and Marina, it's still too dangerous for Cesar to let go. They're still yards from shore, and one misstep will plunge them back into the swift water. Cesar's too exhausted and battered to endure another round of that. Other hikers now dot the shore; they're unable to reach him, but they call out to him, encouraging him to hang on.

These bystanders also share a bit of good news: Abigail and eight-year-old Acis are safe. Cesar breathes a small sigh of relief and looks down at Marina. "Hang on, baby girl," he whispers. "We'll be out of here soon."

* * *

All fourteen members of the Garcia family were within fifteen feet of each other when the flood slammed into them. Abigail managed to grab a tree branch and pull herself to a ledge, where she waited out the flood. Acis was closer to the shore and was able to swim to a rock, which he clung to until a stranger yanked him out. Cesar, with Marina, clung to the tree for two more hours, until the water calmed enough for rescue crews to reach the father and daughter. Marina was completely uninjured; Cesar suffered cuts to his legs from the debris.

Had he been unable to grab that second branch, Cesar and Marina would've been swept over a waterfall a few dozen yards downstream.

Tragically, the remaining ten members of the Garcia party

all perished. They included matriarch Selia, 57; Maria, 26, her husband Hector, 27, and their three children, Daniel, 7, Mia, 5, and Emily, 3; Maribel, 24, and her daughter, Erika, 2; Javier, 19; and Jonatan, 13. Most suffered severe lacerations and head trauma; many had their lungs full of blackened ash water.

When our bodies are deprived of oxygen, even for a minute, hypoxia starts. "This means there's insufficient oxygen reaching your tissues," says Cummins. "As your oxygen levels drop, you may feel like you can't think properly. That air hunger starts and gets intense very quickly. It won't take as long as you think before your body forces you to reflexively inhale."

When you reflexively inhale underwater, you're bringing water into the lungs. "There's even less air, so the hypoxia is worsening," says Cummins. "The longer our tissues and organs don't get oxygen, the faster they die. But it'll take about two minutes underwater before you lose consciousness. As you drown, you'll be gagging and choking, but unable to get air, so the panic must be very intense." After you pass out, it'll be another four or five minutes before you die, adds Cummins.

Near-drowning victims sometimes describe this panic being followed by a moment of euphoria and zen, just before being saved. "We don't know for certain," says Cummins, "but right before you die, your brain may adjust, and you may be completely at peace."

HOW TO SURVIVE A FLASH FLOOD

As the Garcias experienced, there's little warning before a flash flood develops. Flash floods result after excessive rainfall in a short span of time—whereas regular floods can last for days or weeks. Knowing what to do before and during a flash flood can save your life.

BE PREPARED. If you're in an area prone to flash flooding, know where safe areas of refuge are. Gather gear, including flashlights, tarps, a first-aid kit, rain gear, and a portable radio.

SEEK HIGHER GROUND. When a flash flood hits, move to higher ground as fast as you can. If you're able to, alert authorities with your location and a description of your surroundings.

DON'T WALK IN WATER. Swift-flowing water is deadly, and most people misjudge water depth when on roads. Water more than six inches deep can knock you off your feet.

AVOID HAZARDS. Debris will be everywhere, and it can be anything from small tree branches to cars and large boulders to entire houses. Position yourself in an area that will protect you from moving objects. Downed power lines can be dangerous too, so make sure you're not under any utility poles.

PROBE THE GROUND. If you have to walk, don't step in any running water. Use a stick or branch to poke the ground in front of you. What you may believe to be a level road could be concealing cracks or sinkholes. The dirt below the road could erode below the surface, so make sure that where you're about to step can hold your weight.

DON'T DRIVE INTO WATER. If you're in a car and the road ahead of you is flooded, you won't know the depth of the water covering the road. Water more than twelve inches deep can float most cars and trucks. Don't risk it. If you encounter water on the road, turn around and do not drive into it.

LEAVE A TRAPPED CAR. If floodwaters overtake your vehicle, before the flood depths reach five or six inches—the height where walking is no longer recommended—leave the vehicle and seek higher ground.

OPEN WINDOWS OF FLOATING CARS. If it's unsafe to leave your vehicle on foot and the vehicle is taken by floodwaters, immediately unbuckle your seatbelt and open your window. Smash it if necessary. Once the doors of your car are below the waterline, you won't be able to open them, so you need to crawl out of the window and get to the roof of the car.

DO NOT LEAVE A FLOATING CAR. If your car is being carried away by the water, some survival experts suggest swimming free of the car. That's bad advice; survival rates are higher for people who stay with the vehicle as it moves in a flood. It's easier to spot a car than a person, and you have a buffer against larger debris. Lie on the roof for stability, but don't strap yourself to the vehicle. If it rolls over, you'll need to be able to move freely. Call 911 if you can.

SWIM PERPENDICULAR. If you're swept up by the flood, don't try standing. Your foot may get caught in a storm drain or hole in the terrain and you'll be pinned under the current. Instead, if you're able to, swim perpendicular to the current until you get to safety. Given how fast the floodwaters may be, though, swimming may be impossible.

SURVIVAL 101:

NATURAL DISASTERS

Earthquakes, avalanches, and flash floods upend our world in mere seconds, and they can set off a cascading series of life-threatening issues. These five tips can help you avoid injury or death.

1. HAVE A PLAN

If you live in—or are traveling to—an area prone to natural disasters, preparation goes a long way. Create a plan to navigate your actions following the emergency. This can include knowing where the nearest storm centers are located and determining meeting points for loved ones if you become separated. Always heed local disaster warnings.

2. ACT QUICKLY

Natural disasters strike without warning. The instant one begins, you need to get moving toward a safe place. Stalling or hesitating, even for a few moments, can be the difference between life and death.

3. SHELTER SAFELY

Understand the best options for sheltering from any type of disaster and make sure you can reach those places quickly. Whether that's a sturdy table to dive under during an earthquake or knowing to climb on top of your vehicle when trapped in a flash flood, get to the safest spot you can.

4. BREATHE DEEPLY

To prevent shock from taking over and immobilizing you, breath work can help keep you in the moment. You want to breathe deeply, with longer exhales than inhales, to allow your frontal lobe to come back online, enabling higher-level thinking and planning.

5. PROTECT YOUR HEAD AND NECK

In many of these situations, there will be large pieces of debris flying around you at high velocities. If you're unable to get to shelter, cover your head and neck with your hands and arms to avoid blunt trauma to those areas, which can instantly incapacitate or kill you.

EPILOGUE

Surviving the most extreme and intense scenarios boils down to a few basic truths:

Being well prepared for any situation will increase your chances of living through any mishaps that arise.

Those who are able to take a beat to regain control of their body and mind fare far better. Use breath work to re-engage the frontal lobe, enabling more complex thinking and reasoning.

Your brain is arguably the most powerful tool you have during times of crisis. From positive thinking to dissociation to shock to releasing hormones, endorphins, and other neurotransmitters in milliseconds, your mind will do everything it can to keep you alive.

Your body's thresholds for pain are far greater than you realize. The sensation of pain exists to warn us when things are happening to our bodies that may cause harm. In life-or-death moments, we learn that our true pain limits are significantly higher than we ever believed.

While durable and resilient, our bodies function optimally within narrow windows of tolerance. Small things, like temperature fluctuations or foreign bacteria, can disrupt our homeostasis and set off a chain reaction of events that threaten our health. Being properly equipped with gear to combat excessive heat and cold is a must for any survival situation.

Last, you need a bit of luck. Preparation, physical fitness, and situational training is important, but sometimes it's simply a matter of millimeters that keep your femoral artery away from a shark's razor-like tooth or your head from a crumbling concrete wall.

If you take nothing else away from these breathtaking stories and the accompanying expert insights into our physiology, let it be this: When pushed to the limits, the human body is wonderfully engineered to respond to the best of its ability—and survive.

You're built to beat the odds.

ACKNOWLEDGMENTS

FROM SEAN: A massive thank-you to Brandi Bowles, at United Talent Agency, for recommending me to author this book. And thanks to the fine folks at Wondery, including Nicole Blake, Jenny Lower Beckman, and Andy Hermann, for entrusting me to translate their wonderful podcast into the printed realm. Andy, in particular, made hitting our tight deadlines and vast deliverables manageable and easy. Steve Fennessy's insightful edits and questions during initial drafts helped tighten and heighten the book, and for that I'm grateful. And Andrew Yackira, Mauro DiPreta, and the team at William Morrow were wonderful collaborators with editing and shaping the book into what you see on the page now.

To the medical experts featured—Dr. Matt Cummins, Dr. Anthony Giovanone, Dr. Beth Palmisano, Dr. Alex Sabo, Dr. Deepak Sachdeva, and Dr. Olga Terechin—I am immensely thankful for your participation. I appreciate your taking time from busy medical practices and emergency room shifts to answer every question I had, patiently explaining dense biological concepts until I could understand them.

David Meisels, thanks for the late-night laughs when I needed a break from writing and listening to me talk through every chapter. Mike LaMarche, thanks for being a sounding board and your encouragement when things seemed overwhelming.

Joanna Molloy and George Rush, my favorite editors and second parents, thank you for hiring me nearly two decades ago. Everything meaningful that I learned about writing and interviewing happened when I worked for you, and I am forever indebted.

Thanks to my family, including my parents, Diane and Fran, for reading everything I write, even the early clumsy high-school newspaper clippings. To my wife, Rashna, your unending support is vital to my success. You always push me forward, even when I can't see the path. To my little Weller, you have an innate sense of when to burst into my office to give me random hugs. I treasure those. Love you, bubs.

FROM WONDERY: We'd like to thank the whole team behind the *Against the Odds* podcast, particularly our hosts, Mike Corey and Cassie De Pecol; our production team, especially Andy Hermann, Aleta Rozanski, Emily Frost, Matt Gant, and Dezi Blalock; and all the writers, story editors, fact-checkers, and sound designers who bring the show to life each week. Special thanks also to Bryan Kluge for his leadership on the book's visuals; to our publicist, Alice Zou; and to our brand and franchise team, Elias Warren, Greg Salter, and Jill Tully.

ABOUT THE AUTHOR

SEAN EVANS is a seasoned journalist and editorial leader. As a journalist, Sean has written or edited for *Fast Company*, *GQ*, *New York*, *Robb Report*, *Conde Nast Traveler*, *Men's Journal*, *Gear Patrol*, and *Sports Illustrated*, among others. Previously, he was the digital director of *Men's Health*, where his fascination with physiology began.

No stranger to the survival nature of this book, Sean's beaten the odds twice. After an all-terrain vehicle he was riding rolled over on him in the Kalahari Desert in southern Africa, his arm snapped in half, six ribs and his wrist were broken, and his lung was deflated. It took hours to reach a hospital for emergency surgery. Sean was also a passenger in a supercar when it drove off a California mountain at 70 mph, traveling three hundred feet out and eighty feet down. Thankfully, there were no injuries.

Sean resides in Maplewood, New Jersey, just outside New York City, with his wife, Rashna, his daughter, Weller, and Edgar, the family's geriatric chihuahua.

CREDITS

Illustrations on pages x, 1, 2, 7, 13, 17, 20, 22–24, 27, 35, 39, 44, 47, 56, 60, 65–66, 69–70, 78, 81, 88, 92–93, 96, 98, 106–7, 110–12, 116, 126, 130, 139, 146, 149, 155, 158–60, 166, 175, 178, 180, 186–87, 194–96, 199, 201, 210, 214, 217–18, 228–29, 232, 239, 246, 248, 250–51, 254, 262, 264, 268, 270, and 277–79 © Mary E. Pagone.

Stock illustrations: pages i and 160, rattlesnake © Morphart Creation /shutterstock.com; page ii, binoculars © Design_Stock7/shutterstock.com; page ii, canteen © theerakit/shutterstock.com, crocodile © yod 67 /shutterstock.com, and flame © Kair/shutterstock.com; page iii, tornado © jenesesimere/stock.adobe.com; page 21, raft © pandavector/stock.adobe .com; page 24, pattern © armo.rs/shutterstock.com; page 43, solar water still © Pepermpron/shutterstock.com; pages 44 and 188–89, mountain © phipatbig /shutterstock.com; page 66, jungle © Leavector/shutterstock.com; page 88, snow © Tartila/stock.adobe.com; page 104, sunburst © KY/stock.adobe.com; page 105, polar bear @ jenesesimre/stock.adobe.com; pages 108–9, 192–93, and 284–85, pattern © kosmofish/shutterstock.com; pages 44 and 108, mountain © Net Vector/shutterstock.com; pages 112 and 124, sharks © aksol/stock.adobe .com; pages 126 and 143, bear © Turaev/stock.adobe.com; pages 146 and 192, crocodile © kuco/stock.adobe.com; page 172, IV bag © pandavector/stock .adobe.com; page 177, rattlesnake © Morphart/stock.adobe.com; page 178, wolf © Arthur Balitskii/stock.adobe.com; page 188, wolf © Aliaksandr Radzko /stock.adobe.com; pages 188–89, mountain ©phipatbig/shutterstock.com; page 196, circles © Social Media HUB/shutterstock.com; page 213, ground crack © klyaksun©stock.adobe.com; page 232, fire pattern © riansa28 /shutterstock.com; and page 254, spiral © Net Vector/shuterstock.com.

A quick note about the stories in this book: They're inspired by true events, but some elements including dialogue may be invented. Everything is based on research, including what we know about the individuals from their own accounts or the accounts of authors or journalists who reported on the stories. The content of the stories is not intended to be used as a survival manual and is not a substitute for professional medical advice, diagnosis, or treatment.

HarperCollins books may be purchased for educational, business, or sales promotional use. For information, please email the Special Markets Department at SPsales@harpercollins.com.

FIRST EDITION

Designed by Leah Carlson-Stanisic

Library of Congress Cataloging-in-Publication Data has been applied for.

ISBN 978-0-06-338716-4

25 26 27 28 29 LBC 5 4 3 2 1